AGROECOLOGY

Edited by **Vytautas Pilipavičius**

Agroecology
http://dx.doi.org/10.5772/58736
Edited by Vytautas Pilipavičius

Contributors

Martin H. Bremer-Bremer, M. A. Martin-Luengo, M. Yates, Kyoung Jae Lim, Osman Mohammed Babikir, Solomon Muchina, Ameha Sebsibe, Adan Bika, Agol Kwai, Caroline Kirungu, George Obhai, Samuel Wakhusama, Thomas Hilger, Carlos Chávez, Carlos Fuentes, Fernando Brambila, Nami Morales-Durán, Suren Kulshreshtha, Daniel G Bucur

CBS, Edition **2017**

Published by InTech
Janeza Trdine 9, 51000 Rijeka, Croatia

Publishing Process Manager Iva Simcic

Technical Editor InTech DTP team

Cover InTech Design team

Additional hard copies can be obtained from orders@intechopen.com

Agroecology, Edited by Vytautas Pilipavičius

p. cm.

ISBN 978-953-51-2130-5

INTECH

open science | open minds

Contents

Preface

This book is a collection of chapters, concerning the developments within the Agroecology field of study. The book includes scholarly contributions by various authors pertinent to Agricultural and Biological Sciences. Each contribution comes as a separate chapter complete in itself but directly related to the book's topics and objectives.

The book contains 8 chapters:

1. Agricultural Systems in IGAD Region — A Socio- Economic Review

2. Characterization of Industrial Highly Organic Wastewater to Evaluate Its Potential Use as Fertilizer in Irrigation of Agricultural Land

3. Wastes in Building Materials Industry

4. Soil Salinity Control in Irrigated Land with Agricultural Drainage Systems

5. Seeds of Change — Plant Genetic Resources and People's Livelihoods

6. Challenges and Opportunities in Estimating the Value of Goods and Services in Temperate Grasslands — A Case Study of Prairie Grasslands in Manitoba, Canada

7. Modelling of Best Management Practices in Agricultural Areas

8. Effluent Cleaning, Greener Catalysts and Bioecomaterials from Agricultural Wastes

The target audience comprises scholars and specialists in the field.

Prof. Dr. Vytautas Pilipavičius
Aleksandras Stulginskis University
Institute of Agroecosystems and Soil Sciences
Lithuania

Chapter 1

Agricultural Systems in IGAD Region — A Socio-Economic Review

Osman Babikir, Solomon Muchina, Ameha Sebsibe,
Adan Bika, Agol Kwai, Caroline Agosa,
George Obhai and Samuel Wakhusama

Additional information is available at the end of the chapter

http://dx.doi.org/10.5772/60012

1. Introduction

Agriculture is defined to be the management of natural environment in an attempt towards its domestication. The goal of this domestication is to provide humankind with an adequate, controlled, and reliable source of food and fiber. To achieve this goal, agriculture deals with the management of living systems at many scales [1].

The quality and availability of land and water resources, together with important socio-economic and institutional factors is essential for food security [2]. "How to feed the world" is an increasingly urgent rising concern voiced by many people, from local community groups to national and international communities. Agriculture is in crisis. Although the world's agricultural lands continue to produce at least as much food as they have in the past, there are abundant signs that the foundations of their productivity are in danger [3].

Preserving productivity of agricultural land over a long term requires sustainable food production. This could be possible through alternative agricultural practices with considera-tion to social, cultural, political and economic systems [4]. High productivity levels usually come at high environmental and social costs when farmers along the globe practiced the so called conventional agriculture which is the type of farming where some technological advancement is used to gain those high levels. This use of advanced techniques, according to Gliessman [3] is based on science and research (fertilizers, new varieties, irrigation techni-ques,), but this happened at the expense of degrading the basis of natural resources which are the key pillars of agricultural production. These natural resources like water, soil and natural genetic diversity have been undermined by the current technological advancement and in addition to their degradation, there is also the dependence on nonrenewable fossil fuels and help to forge a system that increasingly takes the responsibility of growing food out of the hands of farmers and farm workers who are in the best position to be stewards of agricultural lands [5]. Factory-farm livestock production is another manifestation of the specialized trend in agriculture. The rise in factory farming is coupled with a world-wide trend toward diets higher in meat and animal products. As demand for meat increases, industrialized methods for animal food production become more profitable and wider spread, replacing more sustainable pastoral and mixed crop-livestock systems [3]. On the other hand, with the development of research on natural resources, the term of eco-efficiency originates. Eco-efficient agriculture means increases productivity while reducing environmental impacts. It meets economic, social and environmental needs of the rural poor by being profitable, competitive, sustainable and resilient. The increased food insecurity and vulnerability of a large number of people worldwide point to a broken food production and distribution system. We need to look at the contribution agriculture should make not only to feed a growing population but also to impact less on the planet's resources. The future food supply equation needs to consider the current reality of lower growth rates for major crop yields in conventional agriculture, eco-efficient approaches to diminish impacts on natural resources, the climate change challenge and the volatility of energy prices [6]. This implies that adoption of eco-efficient practices, approaches and eco-efficient farming systems will surely lead to higher productivity levels while maintaining lesser negative environmental impacts. More clearly, Koohafkan *et al.* [7] reported that, there are many competing views on how to achieve new models of a bio-diverse, resilient, productive, and resource-efficient agriculture that humanity desperately needs in the immediate future. Conservation agriculture, sustainable intensification production, transgenic crops, organic agriculture, and agro-ecological systems are some of the proposed approaches, each claiming to serve as the durable foundation for a sustainable food production strategy [7].

Therefore, one of the current widely used ideas about food systems is what is called by Francis et al., (2003) the use of integrating approach that combines ecology of these food systems with the economic and social dimensions [8]. Hence, agroecology has been defined as linking ecology, culture, economics, and society to sustain agricultural production, healthy environments, and viable food and farming communities [9]. It has been reported that, socio-economic, technological and ecological components constantly interact, creating a complex feedback mechanism that through time has selected for the type of food production systems that we observe today [10]. According to Franci *et al.* [8], agro-ecolgy is defined as the study of the whole food system, embracing both natural and social sciences, and emphasizing systems thinking and ecological principles [8].

In Africa, Andriesse et al., [11] has emphasized that agricultural productivity must be increased to meet the demands of an increasingly urban population, as much as to support sustainable rural population [11]. Most of Africa's poor are rural, and most rely largely on agriculture for their livelihoods. The now widely-shared view is that improving agriculture, particularly smallholder agriculture, is fundamental to overcoming the seemingly intractable problem of African poverty. But how? During the past decade Africa has also experienced several episodes of acute food insecurity, with tragic loss of lives and livelihoods. Recently, the Sahelian and the Horn of Africa encountered yet another food crisis that has severely affected millions of people. Droughts, crop failures and other disasters often trigger these crises. But the real causes go deeper and they are diverse [12].

In Africa in general, there is a growing debate based on the fact that population growth has exceeded the carrying capacity of land at the current technological levels, which will have bad consequences such as environmental and ecological deterioration, wide spread poverty, malnutrition and famine. As well, in some countries this may lead to conflicts and political instability. The horn of Africa or what is also known as IGAD region is not an exception from that situation, if not is more severe and deteriorated. It has been emphasized by Giessen [13] that, in the Horn of Africa increasing scarcity and degradation of natural resources seriously threatens human well-being. The population in the region has increased fourfold in the past 50 years and continues to go rapidly. Farmers need to feed more mouths and extra areas of land are needed, at the expense of forest and pasture lands. With a high and stable number of pastoral communities and decreasing amount of pasture land, pressure on land and water grows. The mounting needs for fertile soils and irrigated land is intensified by high international demands for food and energy [13].

2. Conceptualizing farming systems

Farming systems, and ways of thinking about them, evolved in space and time. Rapid evolution took place in the last two decades when crop and livestock yields increased, together with concerns about their socio-economic and biophysical tradeoffs [14].

Systems in all sectors of the society, including agriculture need to be examined through the system approach. The term system can imply different things such as a process, procedure or unit [14].

Generally, systems could be classified into natural, social and artificial systems. Natural systems are those that exist in nature. Such an apparent example of these natural systems on which agriculture depends, is rock weathering to form soil, plants sustained by such soil; animals sustained by such plants. The second type of systems is the social form which essentially consists of the entities forming animate population, the institutions or social mechanisms created by such entities, and the interrelationships among/between individuals, groups, communities, expressed directly or through the medium of institutions. The third type is the artificial systems, which are created by humans to serve their purposes. Thus agricultural systems are examples of such artificial systems which are directly dependent on either or both natural and social systems, or indirectly on natural systems through the dependence of social systems themselves on natural systems [15].

Any farm as a unit could be a factory for decision making, it can be a production unit for either crops or livestock or a mixture from both of them. The farmer and other human elements of the farm, the physical and biological environment are the boundaries where this farm as a system operate, and it may change, so it is dynamic. So as pointed out by Dixon *et al.*[16] farming system approach considers both biophysical dimensions and socio-economic aspects at the level of the farm, where most of the agricultural production and consumption decisions are taken. The power of this approach lies in its ability to integrate multidisciplinary analysis of production and its relationships to the key biophysical and socioeconomic determinants [16]. Between the constituents of any farming system, the human, the physical and biological parts, there is complicated interactions between so many detailed components. For example, the human element may be a set of household members including family labour, which in addition to hired labour constitute the multi-nature of each constituent. Also the soil is not only such particles where the plant grow, but a series of physical and chemical characteristics and reactions, all of which are very important for the plant life cycle.

Generally in the literature, there are so many definitions to farm or farming system, each one of them was trying to define the term from different perspective: Okigbo [17] defined a farming system as an enterprise or business in which sets of inputs or resources are uniquely orchestrated by the farmer in such a way as to satisfy needs and to achieve desired objectives in a given environmental setting. It could also be defined as a decision making unit as it transforms land, capital, and knowledge into useful products that can be consumed and sold [18]. According to McConnell and Dillon, [15], the term farming system refers to the cultivation patterns used in a plot conceptualized in relation to the farm, other agricultural entities, the socioeconomic and ecological context and the technology available that determine its character [15]. This implies that a farming system is a part of a larger ecological, social, political, economic, cultural environment that is affecting its characteristics. Hence, it is clear from the definitions that farming systems or agro-ecosystems are comprised of many components and agents. The components could be biophysical, socio-economical, and cultural and the agents could be species, ecosystems, households, social communities, scientists, policy makers. Those components and agents are operating on different scales (e.g. local, national, global) while pursuing different objectives.

Rosen [19] defined life as an open process of autopoiesis distinct from the type of external driven organization typical of machines. So building on this definition, Gomeiro et al, [1] mentioned that agriculture implying dealing with life and agricultural systems are also agro-ecosystems, and agricultural science can be referred to as agro-ecology. Furthermore, Altieri, [20] defined agro-ecosystems as communities of plants and animals interacting with their chemical and physical environments that have been modified by people to produce food, fiber, fuel, and other products for human consumption and processing [20]. In this regard, and as pointed out by Kerr, [21] farms can be considered as ecosystems managed by farmers; thus agriculture is concerned with farmer-managed ecosystems.

Norman and Malton [22] distinguished four main phases in the evolution and development of farming systems approach:

- Predetermined focus, for instance on improving cropping system. Emphasis was on normative and prescriptive issues through application of techniques such as budgeting, linear programming and other tools for decision making analysis.
- Whole farm focus, with the contribution of farm management studies.
- Natural resource focus due to conflicting interests between strategies designed to improve short-run productivity and long-run ecological sustainability.
- Sustainable livelihood focus, which includes a wider set of issues, not just production. The concept emerged nearly simultaneously in the farming system literature and in a series of international conferences.

3. Characteristics of farming systems

Agricultural systems, even the most traditional ones, are not static systems; in fact they are dynamic [7].

Spedding [23] emphasized that the classification of agricultural systems has a long history, but there is no generic system that is truly comprehensive and can serve all purposes [23]. They exhibit great diversity and have been classified in various ways including an ecologically based classification [24], [25]. According to Fresco and Westphal [25] there are basically two ways to classify farm systems. First the farm systems of the world can be grouped together in broad classes that reflect fundamental structural differences, for example, plantation systems, tillage system (with and without livestock), alternating systems and grassland systems [26]. The second approach is that used by Grigg [27] who makes explicit reference to geographical units. These classifications and others have in common that they combine economic and biological factors. The main usefulness of this type of broad classification lies in its indication of the relative importance of different classes of farm system and their relevance to the setting of priorities in international agricultural research. The weakness of these past attempts is that they provide little systematic insight into the way the classification relates to the development of agricultural technology. Furthermore, all these approaches classify elements of farm systems (livestock, crop, capital use) but do not do justice to the interaction of the elements which make up the system [25].

Existing classification are based on a wide variety of factors and differ markedly in their utility, comprehensiveness, and ability to be mapped [28]. A summary of comparison between the existing global classification systems is illustrated in the table 1.

Classification	Crops status	Livestock status	Categories no.	Pros and cons
Ruthenberg 1980	- Degree of cultivation - Forest, bush, savannah, grass - crop type - Irrigated versus rainfed	Degree of movement/ permanence	8 major	Categories too broad and incomplete
Grigg 1972	- crop type -commercialization - location/agro-ecology	Degree of movement/ permanence	9 major	System incomplete and somewhat selective '
Dixon et al 2001	- crop type - commercialization - location/agro-ecology	Degree of movement/ permanence	8 major 72 globally by region	Derivation not explicit, difficult to map using existing global data set
Sere and Steinfeld 1996	- Are there crops or not? - Rainfed versus irrigated - Agro-ecology	- Landless or rangeland based - Agro-ecology	11 major	Livestock based, so no categorization of crop systems, can be mapped using appropriate proxies
Explicit AEZ method, e.g. Fischer et al. 2002	Match land suitability to crop requirements for given inputs and technology	Not dealt with though probably could be included	As required	Easily mapped, assesses what may be, rather than what actually is
Statistical classification, e.g. Wint et al. 1997	Cluster spatial units based on crop densities, intensities	Cluster spatial units based on livestock densities	As required	Easily mapped, arbitrary, data sensitive, and non-replicable

Source: adapted from Robinson et al., [28]

Table 1. Comparison between existing Global Livestock Classification Systems

A system is characterized by its elements, their inter-relationships and by definition of the boundary of the system. It could also be open, in a sense that external relationships are also included. However, systems at each level are inter-linked and even with sub-systems [29].

Around the world, agricultural ecosystems show tremendous variation in structure and functions, because they were designed by diverse cultures and diverse socioeconomic conditions in diverse climatic conditions [30]. According to John Dixon et al, [16] the following is the key biophysical and socioeconomic determinants of a farming system:

1. Natural resources and climate
2. Science and technology
3. Trade liberalization and market development
4. Policies, institutions and public goods
5. Information and human capital

These categories represent the major areas in which farming system characteristics, performance and evolution are likely to be significantly affected over the next thirty years. Some of these factors are internal to, or part of the farming system, whereas others are external. Policies, institutions, public goods, markets and information are external and they influence the development of the farming system. Technologies which determine the nature of production and processing, and natural resources, are largely endogenous (internal) factors. In general terms, the biophysical factors tend to define the set of possible farming system, whilst the socio-economic factors determine the actual farming system which can be observed at a given time [16]. In the African context, for example, Guyer and Peters [31] mentioned that there is an extensive literature on African agrarian systems that highlight how social and cultural relations shape agricultural production and investment, the type of technologies adopted, and the operation of agricultural markets.

Each individual farm or farm system has its own specific characteristics arising from variations in resource endowments and family circumstances within the context of local institutions and policies. These are translated into productive activities, and household consumption and decision making activities. In the context of sustainability, Koohafkan, *et al.* [7] had suggested, based on extensive literature review, a series of attributes that any agricultural system should exhibit in order to be considered sustainable, the following are these basic attributes:

a. Use of local and improved crop varieties and livestock to enhance genetic diversity and adaptation to changing biotic and environmental conditions.

b. Avoid the unnecessary use of agrochemical and other technologies that adversely impact on the environment and human health.

c. Efficient and reduced use of resources, nonrenewable energy and farmer dependence on external inputs.

d. Harness agro-ecological principles and processes.

e. Making productive use of human and social capital to enhance solidarity and exchange of innovations and technologies.

f. Reduce the ecological footprint of production, distribution and consumption practices.

g. Promoting climate adaptive practices.

h. Enhanced adaptive capacity to strengthen the ability to adequately respond to changes.

i. Strengthen adaptive capacity and resilience of the farming system by maintaining agro-ecosystem diversity.

j. Recognition and dynamic conservation of agricultural heritage systems.

The design of agro-ecosystems that exhibit many of the attributes of sustainability has become a leading objective of scientific research and policy agendas [7].

4. IGAD region: A general background

The easternmost part of the African continent is often referred to as the Horn of Africa. Some geographers considered the Horn of Africa is comprised of Ethiopia, Eritrea, Somalia and Djibouti. Others include to it Sudan, Kenya and Uganda. Then due to the complexities in defining the region, some organizations apply the term Great Horn of Africa. Major International Organizations including European Commission and the Intergovernmental Authority on Development (IGAD) use the term Horn of Africa to refer to the states of Djibouti, Eritrea, Ethiopia, Kenya, Somalia, South Sudan, Sudan and Uganda [13].

Figure 1. A map showing IGAD region

IGAD region or the so called horn of Africa according to Mengisteab [32] has an estimated total population of about 226.9 million in 2012 and a total area of 5,209,975 sq km the countries of the region are all members of the Intergovernmental Authority on Development (IGAD), although Eritrea's membership in the regional body has been suspended since 2007, but now the country has applied for reinstatement. Two of the youngest countries of the region, Eritrea and South Sudan, were formed through secessions from Ethiopia in 1993 and from Sudan in 2011, respectively.

Tekle [33] has emphasized the issue of diversity and the fact that the countries of the region share certain characteristics. Among these characteristics he mentioned that their societies are divided along ethnic and religious lines, with political loyalities cut across state boundaries [33]. When recalling the pre-colonial history of the region, it had complex and diversified political structures or systems, and the societies were dynamic or there were changes or continuities [34]. One of the notable characteristics, according to Mengisteab, [32] the region is a mosaic of cultures with considerable ethnic diversity both regionally and within countries. If language can serve as a proxy for ethnic identity, the region is said to be home to some 340 languages. Lewis [35] reported that, Sudan (both north and south) is said to have 134 languages, followed by Ethiopia with 89 languages, Kenya with 62, Uganda with 43, Eritrea with 9 and Djibouti with two local languages [35]. The countries of the region are also characterized by religious diversity with various denominations of Christianity and Islam coexisting, along with various forms of traditional religion. Many of the region's ethnic groups are also split across several countries by national boundaries established by colonialism. The Somali people, for example, live in Somalia, Djibouti, Ethiopia and Kenya. The Beja, Tigre and Rashiada live in Sudan and Eritrea. The Tigrigna, the Kunama and Shaho live in Eritrea and northern Ethiopia; the Oromo live in Ethiopia and Kenya, the Afar live in Eritrea, Ethiopia and Djibouti. The Luo are spread over Kenya, Ethiopia, South Sudan, Uganda, Tanzania and Eastern Congo, while the Luhya live in Kenya, Uganda and Tanzania [32]. Michalopoulas and Elias [36] examined the impact of ethnic portioning on civil conflict and economic development in Africa. They found that, out of a total of 834 ethnicities, for 231 ethnic groups at least 10% of their historical homeland falls into more than one contemporary states. Then with a threshold of 20%, about 164 ethnicities have been portioned across the borders. Using regional data on civil wars in Africa (1970-2005), they found that, civil conflict is concentrated in the historical homelands of partitioned ethnicities. Also border areas populated by ethnic groups only modestly affected by the artificial border design also experience more conflict. Examining the effect of ethnic portioning on economic development, showed that development in the historical homeland of portioned ethnic groups is lower by almost a half, compared to non-portioned ethnic areas. However, this ethnic portioning was resulted from the pre-colonial artificial drawing of African political boundaries among European powers in the end of the 19th which led to the portioning of several ethnicities across African states [36]. The partition of ethnic groups into different countries often involves the disruption of social and cultural ties [32]. In the case of pastoral communities partition also implies disruption of economic process as it hinders the movements of groups who rely on regional ecosystems for survival [37]. However, despite these ethnical portioning and conflicts along the borders and borderlands, there is always hope that these borders and borderlands could be used as potential resources rather than triggering points of conflicts. Borders in Africa have generally been conceived as barriers [38]. But they have also been conceived as 'conduits of opportunities' [39]. Frequently, Horn of Africa is associated with natural and man-made catastrophes, which often have a cross-border dimension, and with violent border conflicts. Many local people as well as external observers perceive the arbitrary colonial borders as one of the causes for these conflicts. Four different types of resources could be extracted from state borders and borderlands. These are: first economic resources (cross border trade and smuggling), second, political resources (access to alternative centers of political power, trans-border political mobilization), third, identity resources (as security device in an inter-ethnic competition) and fourth, status and rights resources (citizenship and refugee status) [38].

Addressing this challenge, IGAD Centre for Pastoral Areas and Livestock Development (ICPALD) is now implementing an intervention that will solve this problem. This intervention will include adopting a regional protocol for transhumance mobility which will enable such groups to maintain their economic, social and cultural ties across national boundaries. Such arrangements, according to Mengisteab [32], if absent will become a source of instability and major conflicts.

Another characteristic of the Horn of Africa countries is the dichotomy of modes of production that govern their economies. The modes of production operating in the region range from a capitalist sector symbolized by emerging stock markets and relatively advanced financial systems to subsistence farming and pastoral economic systems, which are essentially non-capitalistic [32].

Many people perceive the Horn of Africa as a doomed desert area, with clashes between tribes over food, cattle and water. Less people perceive it as a region endowed with great natural resources, an extraordinary variety of flora and fauna, and powerful people with very rich cultures, who have also proved to be well capable of realizing firm economic development [13]. Table 2 presents some characteristics of the region.

Item	Djibouti	Eritrea	Ethiopia	Kenya	Somalia	South Sudan	Sudan	Uganda
Area (sq. km)	23,200	117,600	1,105,300	580,367	637,657	640,000	1,886,068	241,038
Population in 2013('000)	873	6333	94101	44354	10496	11296	37964	37579
Population growth (1970-1990)	6.2	2.7	2.6	3.7	3.0		2.9	3.1
population growth (1990-2010)	2.3	2.5	2.7	2.7	1.7		2.5	3.2
Projected population growth (2010-2030)	1.8	2.3	1.8	2.4	2.8	2.5	2.1	2.9
Population in 1960	85.0	1,424.0	23977.0	8105.0	2,819.0		11,562.0	6,788.0
Population projections for 2030	1,263	8,394	118,515	65,928	16,360	15,082	51,775	55,846.0

Source: Adapted from Mengisteab [32] and African Statistical Yearbook [40]

Table 2. Area and size of population of the countries of IGAD region

4.1. Socio-economic conditions of IGAD region

In the Horn of Africa, three out of four people reside in rural areas, and rely on subsistence production. Economic growth in the region averaged about 3% between 1965 and 1995, with a 3% population growth that led to a stagnation of per capita income at US$ 223 at the end of the period compared to US$ 491 in the rest of Sub-Saharan Africa. Since the second half of 1990's, most countries in the region experienced strong growth. The region grew by 4.4% between 1996 and 2000, and at 5.3% and 7.9% in the first and second half of 2000's respectively. The overall disappointing growth performance in the Horn of Africa has led to poor socio-economic indicators [41].

The region is considered one of the most politically unstable regions in the world. This vast area is linked by a shared history of conflict and a complex web of economic ties. Economic exchange is also a feature of the region with growing trading links with the global economy. As one of the Africa's eight Regional Economic Communities, the Intergovernmental Authority on Development has institutional responsibility for advancing economic integration in the Horn. However, the economies of the IGAD region possess significant structural obstacles to the attainment of regional economic integration. Poverty and lack of diversification in the economy are root problems. The population is mostly rural with the largest percentage of the workforce engaged in agricultural labour. The livestock sector form a very important part of the economy, with the Horn of Africa supporting one of the largest concentrations of pastoralists people anywhere in the world. In common with most developing economies, the manufacturing sector remains small. The countries depend for their exports on relatively small number of primary commodities in which they compete each other. Historically, this pattern of trade has produced very low levels of formal intra-regional trade. Another disincentive is the relatively low tax base in some countries. There is also the problem of heavy dependence on imports from outside the region. There are also several important non-economic obstacles to the regional integration, including uneven capacities and different types of state, weak institutions, competing institutional frameworks (IGAD, EAC, COMESA, CEN-SAD) and regional conflicts [42]. Table 3. Presents some socio-economic characteristics of the region.

Item	Djibouti	Eritrea	Ethiopia	Kenya	Somalia	South Sudan	Sudan	Uganda
Total fertility rate (%)	3.4	4.7	4.5	4.7	6.6	8.9	5.4	6.2
rural population (%)	22.8	77.8	81.8	67.7	61.3	82.9	65.2	81.9
Life expectancy at birth (years)	61.8	62.9	59.7	57.7	55.1	59.8	61.8	50.4
Prevalence of undernourishment	19.8	65.4	40.2	30.4	-	-	39.4	34.6
GDP per capita (US$)	1640	543	519	1011	-	1042	1856	624
Human poverty index (%,)	25.6 (2007)	33.7 (2007)	50.9 (2007)	31.4 (2006)	-	-	34.0 (2007)	-
Human Development Index (HDI)	0.467 (170)	0.381 (182)	0.435 (173)	0.535 (147)	0.285 (2012)	-	0.473 (166)	0.484 (164)

Source: African Statistical Yearbook [40], and Human Development Report, 2014 [43]

Table 3. Selected indicators of socio-economic conditions of IGAD region

As mentioned before the region is riddled with conflicts of different types and causes and it is considered one of the hotspot areas in Africa and even at global dimension. However, the region is also facing other severe problems of food insecurity and poverty. The populations of the region have endured many inter-state and intra-state armed conflicts during the post-independence era. The region was by no means peaceful before the era of decolonization. During the period between 1800s and the era of decolonization, for example, it experienced many wars, which revolved mostly around state formation and empire building; slave raids, control of resources, and trade routes; resistance to colonization and the liberation struggle [32]. This implies that, the region is historically well prepared to fall in conflicts, particularly when other factors such as the undefined borders which were considered as time bombs triggering conflicts here and there.

The region's post-independence conflicts can be classified into six categories, including direct and indirect inter-state wars and armed conflicts; cross-border inter-communal conflicts; civil wars and civil conflicts; conflicts among rebel group over differences of political programmes and power struggle; intra-state inter-communal conflicts; and one-sided violence perpetrated upon civilians by the state or other armed political groups [32]. The allocation, use and management of natural resources such as water, fertile land, pasture land, trees but also oil, lead to large internal and international tensions in virtually all countries of the Horn of Africa. Through political tensions and grievances about the loss of livelihoods among farmers, pastoralists and fishermen, regional insecurity is rapidly increasing. [13].

The major economic driver of this region is crop and animal agriculture. However, a large proportion (60-70%) of the landmass in the IGAD region is covered by arid and semi-arid lands (ASALs) and experiences prolonged drought and unpredictable rainfall patterns (less than 400 mm of rainfall annually). The region is also characterized by chronic conflicts at national or cross border levels. These conflicts are often linked to scarceness of natural resources (water, rangelands) and competition over their use.

Agriculture is the core economic sector of the IGAD Member States. It contributes on average about 15% to the national GDPs and employs about 75% of the population in the region. Due to this, the performance of the agricultural sector is the key determinant of annual changes in the poverty and food security levels in the region.

The predominant livelihood system in the region is pastoral and agro-pastoral production. On average, livestock contributes 57% of the agricultural GDP in the IGAD region. This makes livestock a key contributor to the economies of the IGAD Member States. The nomadic lifestyle of the pastoralists involves a constant search for pasture and freshwater resources. Pastoralism cuts across national boundaries, frequently causing conflicts and necessitating a regional approaches to and collective action from the affected governments. Pastoralism as practiced in the IGAD region is both an economic and a social system that is highly dependent on the rearing of livestock. Livestock are core to pastoral livelihoods and pastoral identity and livestock and livestock products contribute significantly as the main sources of food and income in pastoral households. Sheep, goats, cattle, camels and donkeys are the predominant livestock holdings. Despite, the abundance of livestock in the pastoral areas, the pastoralist communities across the region remain among the most marginalized and face common problems of low productivity of (endemic) indigenous livestock breeds, poor physical infrastructure, limited access to markets, lack of appropriate information, poor communication, lack of access to financial capital and limited access to crucial input services to enhance livestock production within their environments and the threat of dwindling pastures for their flocks. The pastoral areas are characterized by recurrent droughts and other natural disasters. The effects of climate change are reflected in the increasing frequency and severity of episodes of drought. During 2011, the most severe drought conditions to have occurred over the last 60 years were recorded in the Horn of Africa with severe consequences and impacts on the human and livestock populations.

4.2. Disasters and resilience in IGAD region

A global report on disaster hotspots [44], hundreds of disasters occur worldwide each year in locations without sufficient local capacity or resources to prevent death and destruction and to support rapid recovery. According to this report, disaster is defined as a serious disruption of the functioning of a community or a society causing widespread human, material, economic or environment losses which exceed the ability of the affected community or society to cope using its own resources.

Globally, climate change is projected to increase the frequency and severity of extreme weather events like droughts and floods. The worst food crisis in 2011 in East Africa has been caused according to Oxfam report [45] by people and policies, as much as by weather patterns. An adequate response to the current crisis must not only meet urgent humanitarian needs, but also address the following problems:

1. To avoid catastrophic levels of global warming through: increasing efforts to limit global temperature rise and increase and mobilize financial resources by the developed countries
2. To improve food security and strengthen climate resilience with focus on: (a) disaster risk reduction, (b) climate change adaptation (c) long-term investment in livelihood protection measures and smallholder food production [45].

4.3. The Intergovernmental Authority on Development (IGAD): Background

The Intergovernmental Authority on Development (IGAD) is the Regional Economic Community (REC) of the States of Djibouti, Eritrea, Ethiopia, Kenya, Somalia, Sudan, South Sudan and Uganda. IGAD is recognized as one of the Regional Economic Community (REC) that is African Union's building blocks for regional and continental social and economic integration. The IGAD region is a large area occupying about 5.2 million Sq. Km and has a population of about 230 million people growing at a rate of 2.6%. This population constitutes about 20% of the African continent's population. This large economic bloc has great strategic and geopolitical significance and prospects for the people of the Eastern African region.

IGAD was established in 1986 as the Inter Governmental Authority on Drought and Development (IGADD) to coordinate the efforts of the member states in combating desertification and promoting efforts to mitigate the effects of drought. Then, in 1996 at an extra-ordinary

Summit, the Heads of State and Government resolved to re-vitalize the Authority and to expand its mandate to cover political as well as economic issues. Consequently, the Authority was re-named the Inter Governmental Authority on Development with new expanded mandate.

In terms of structure, IGAD consists of four principal bodies, with its head quarter in Djibouti. These principal bodies are Assembly of Heads of State and Government, Council of Ministers, Committee of Ambassadors, and the Secretariat. The Secretariat is the executive arm of the Authority and is headed by an Executive Secretary. The Secretariat in addition to the Office of the Executive Secretary has four divisions, namely Agriculture and Environment, Economic Cooperation and Social Development, Political and Humanitarian Affairs, and Administration and Finance Division. To more effectively implementing its expanded mandate particularly at country and community levels, IGAD has established several institutions and specialized centers of excellence, including the IGAD Climate Prediction and Application Centre (ICPAC), Conflict Early Warning and Response Mechanism (CEWARN), Sheikh Technical Veterinary School (STVS) and IGAD Centre for Pastoral Areas and Livestock Development (ICPALD). There are also IGAD programs, including IGAD Capacity Building Program Against Terrorism (ICPAT), IGAD Regional HIV and AIDS Partnership Program (IRAPP), and IGAD Political Offices.

4.4. IGAD region: Country profiles

4.4.1. Djibouti

Djibouti is a small country strategically located in the Horn of Africa to the southern tip of the Red Sea. It covers an area of 23,300 Km^2 and is bordered by Eritrea, Ethiopia and Somalia. This country was created in the 19th century by France after the opening of Suez Canal in 1869. The opening of a rail link with Addis Ababa in 1917 further increased Djibouti's role as a transit station for both passengers and freight. Djibouti gained independence from France in 1977. Today it is still functions as a major port, transit and communication hub for the Horn of Africa [46]. The coastline which plays a major role in Djibouti's economy, is 314 Km long. The majority of the country is defined as desert and consequently the climate is torrid and dry throughout most of the country [47]. From May to September the climate is particularly hot with daily temperatures of approximately 40°C (104° F). Mid October to Mid April is considered to be cooler season with temperatures averaging about 25°C (77°F) in addition to being the time that there is occasional rain. Djibouti has three distinct geographic areas: the coastal plains which are emphasized by white sandy beaches; the volcanic plateau in the southern and central parts of the country; and in the north, the mountain ranges where the elevation can be as high as 2000 metres above the sea level. While most of the terrain is desert, there are some pockets of forest and dense vegetation in the north [48]. The country is divided into five administrative districts: Djibouti, Ali-Sabieh, Dikhil, Tadjoura, and Obock. It represents a country with relative political stability, economic freedom, and modern financial, transport and communication infrastructure in an otherwise underdeveloped region which is subject to recurrent civil unrest and economic uncertainty [46].

The population of Djibouti is estimated to be 923,000 persons, comprised of two main ethnic groups; the Afars (related to tribes in Eastern Ethiopia) and the Issas (related to tribes in northern Somalia). As well as small portion of other mixed ethnical groups. Two thirds of its population lives in the urban sites mainly in the capital town of Djibouti. Djibouti's economy is characterized by extreme duality, as it is divided between a modern outward-looking urban commercial sector and a rural, subsistence-based pastoralist economy which has little access to infrastructure, services and markets. The economy mainly depends on external sources and service sectoris. With its few natural resources and low rainfall, Djibouti has limited possibilities for agricultural production. Activities in the primary sector therefore make a negligible contribution to the national economy although are extremely important at the rural level, where livestock forms the basis of household livelihoods. The secondary sector, industry and manufacturing are poorly developed because of small domestic market, lack of locally available raw materials and a largely untrained labour force. As compared to other Sub-Saharan countries, where economic planning is central with non-liberal economies, Djibouti has no system of central economic planning and the economic structure is highly liberalized. Major industries include water bottling company, animal feed factory, slaughterhouse and dairy products plant. This is in addition to urban development, transport and communications. The private sector has given some official incentives to encourage investment [46]. The nature of dualism in the economy is explained by having a modern economic rent generating economy coexist alongside an informal economy. The informal economy constitutes a major proportion of the economic activities and provides livelihoods for much of the Djibouti population.

Djibouti adopted its first poverty reduction strategic paper in May 2004, with an incidence of extreme monetary poverty rising from 34.5% in 1996 to 42.2% in 2002. Then many efforts were adopted after the implementation of the strategy and there is significant progress reported. Some improvements in the growth rates of the GDP and social indicators were reported over the period 2004 – 2006. Despite this encouraging progress, the evaluation report mentioned that the priority actions included in the poverty reduction strategic paper have not been satisfactorily implemented [50].

According to table 4, Djibouti's Human Development Index (HDI) value for 2013 is 0.467, with a rank of 170 out of 187 countries and territories. The table shows also HDI trends since 1980.

Year	Life expectancy at birth	Expected years of schooling	HDI value
1980	53.6	2.3	NA
1990	56.7	2.7	NA
2000	57.0	2.9	0.412 (2005)
2010	60.3	6.0	0.452
2013	61.8 (+8.2)	6.4 (+4.1)	0.467 (+13.3%)

Source: Human development Report, UNDP, 2014 [43]

Table 4. Djibouti's HDI trends 1980-2013

Djibouti's HDI is below the average of 0.493 of the countries in the low human development group and of 0.682 for countries in the Arab States [43].

As well, Djibouti earns rents from several foreign military bases especially French and American ones, alongside considerable bilateral assistance aimed at fighting endemic poverty [50]. The country is faced by several environmental and biodiversity threats such as poverty, decreasing vegetation, overgrazing, deforestation, increased development around the coastal zones, depletion of mangrove forests, and continuous loss of wildlife habitat [46].

4.4.2. Eritrea

Eritrea is situated along the western coast of the Red Sea and has a total land area of about 124,000 km^2. The climate ranges from hot arid in the lowlands to temperate sub-humid in the highlands [51]. Eritrea is the Italian form of the Greek name *Erythraia,* meaning "red land". With its capital at Asmara, it is bordered by Sudan to the West, Ethiopia in the south and Djibouti in the east. It is a multi-ethnic country, with nine recognized ethnic groups [52].

According to IFAD [51], Eritrea is a country emerging from a dual crisis of war (1998 -2000) and an unprecedented series of drought (2000-2003). It has unable to build its institutions, suffer from financial constraints and has large number of internally displaced persons, returnees and demobilized soldiers due to those crises [51]. Eritrea population is estimated as 6,130,922 persons, and according to the World Bank ranking, it is classified as low income country. Eritrea has faced considerable challenges over the years, including variable climate conditions. This has been compounded by restrictive economic policies, political isolation, a significant decline in remittances and scarcity of foreign exchange.

The real GDP growth for 2013 fell sharply to 1.1% from 7%, the previous year and is projected to increase marginally to 1.9% in 2014. This growth will largely be driven by copper production at the Bisha mine; the start of gold production at the Zara mine project in 2014; and continued exploration activity and investment in the mining sector. In the medium term, Eritrea sees further prospects in oil production, fisheries and tourism [53]. The economy remains largely managed with the government active in most sectors. Despite on-going efforts to promote more private sector participation in the economy, performance remains sub-optimal. The economy continues to suffer from the effects of the border conflicts, the country's vulnerability to external shocks and the persistent foreign exchange shortages, which are fueling macro-economic imbalances and hampering growth [54].

As far as regional integration is important for Eritrea as for other countries as well, it is currently a member of the Common Market for Eastern and Southern Africa (COMESA), the Community for Sahel - Saharan States, New Partnership for Africa's Development (NEPAD), and the IGAD. This is in addition to some bilateral investment agreements. It is also gradually renewing its engagement with the wider international community [53].

According to table 5, Eritrea's Human Development Index (HDI) value for 2013 is 0.381, with a rank of 182 out of 187 countries and territories. The table shows also HDI trends since 1980.

Year	Life expectancy at birth	Expected years of schooling	HDI value
1980	43.3	NA	NA
1990	48.2	NA	NA
2000	56.1	4.1	NA
2010	61.3	4.1	0.373
2013	62.9 (+19.6)	4.1	0.381 (+2.1%)

Source: Human development Report, UNDP, 2014 [43]

Table 5. Eritrea's HDI trends 1980-2013

Eritrea's HDI is below the average of 0.493 of the countries in the low human development group and of 0.502 for countries in Sub-Saharan Africa [43].

4.4.3. *Ethiopia*

In an area of 104300 km^2, Ethiopia is bordered by five countries, namely Sudan, Somalia, Djibouti, Kenya and Eritrea. Its population which represent second largest one in Sub-Saharan Africa after Nigeria, is estimated at 85900 million persons in 2013, of which about 81.8% live in rural areas. Of the total population, about 73.5% are 14 years old or younger, 79.1% portion represent the economically active population in agriculture, and about 81.8% live in rural areas [40].

Ethiopia is culturally and biologically diverse, with a diverse mix of ethnic and linguistic background; and more than 80 ethnic groups, each with its own language, and about 200 dialects, culture and tradition. It has extremely varied topography, characterized by highland complex of mountains and bisected plateau. Then surrounding the highlands is the lowland grazing areas that form a wide apron surrounding the highland massif and part of the Great Rift Valley [55].

Poverty in Ethiopia is wide spread and multifaceted. The proportion of the population below the poverty line is 77.6% in 2005 and the Gini's index is 29.8%. However, now the government has formulated a five year growth and transformation plan (201 -2105) to eradicate poverty. This plan envisages that besides maintaining a fast growing economy, better results will be realized in all sectors. During this plan period special emphasis will be given to agricultural and rural development, industry, rural infrastructure, social and human development, good governance and democratization [56].

According to table 6, Ethiopia's Human Development Index (HDI) value for 2013 is 0.435, with a rank of 173 out of 187 countries and territories. The table shows also HDI trends since 1980.

Year	Life expectancy at birth	Expected years of schooling	HDI value
1980	43.8	3.2	NA
1990	46.9	3.1	NA
2000	52.2	4.3	0.284
2010	61.5	8.2	0.409
2013	63.6 (+19.8)	8.5 (+5.3)	0.435 (+53.2%)

Source: Human development Report, UNDP, 2014 [43]

Table 6. Ethiopia's HDI trends 1980-2013

Ethiopia's HDI is below the average of 0.493 of the countries in the low human development group and of 0.502 for countries in Sub-Saharan Africa [43].

4.4.4. *Kenya*

The republic of Kenya is a country in the African Great Lakes region of East Africa. It lies on the equator with the Indian Ocean to the south-east, Tanzania to the south, Uganda to the west, South Sudan to the north-west, Ethiopia to the north, and Somalia to the north-east. It covers about 581,509 km^2 and has a population of 44,354,000 persons in 2013 according to the African statistical Yearbook [40]. Kenya has a warm, humid, climate along its Indian Ocean coastline, with wildlife rich savannah grasslands inlands towards the capital. Nairobi, the capital has a cool climate which becomes colder towards Mount Kenya. Further inland there is a warm and humid climate around Lake Victoria, and temperate forested and hilly areas in the western region. The long northeastern regions along the border with Somalia and Ethiopia are arid and semi-arid areas. Lake Victoria, the world's second largest fresh water lake, is situated to the southwest and is shared with Uganda and Tanzania. Kenya, along with Uganda and Tanzania is famous for its safaris and diverse wildlife reserves and national parks [57].

Topographically, Kenya rises from a low coastal plain on the Indian Ocean in a series of plateaus to more than 3000 metres in the centre of the country. All inland regions of semi-arid, bush covered plains constitute most of the country's land area (nearly 70%-80%). In the northwest there is Lake Turkana and Kulal Mountains. In the southwest lie the fertile grasslands and forest of the Kenya Highlands, one of the most successful agricultural production regions on Africa. North of Nairobi, the Kenya Highlands is bisected by the Great Rift Valley, an irregular depression that cuts through western Kenya from north to south in two branches [58].

Population in Kenya, is 41800, about 39.6% aged 14 years and less, 49.5% aged in a range of 15-64 years, while the rest represents 65+ years group. Of the total population, about 67.7% live in rural areas [40]. People of African descent make up about 97% of the population in Kenya and they are divided into 40 ethnic groups. The Kikuyu who make up to 22% of the population constitute Kenya's largest ethnic group. The next largest groups are Luhya (14%), Luo (13%), Kalengin (12%) and Kamba (11%). Small numbers of people of Indian, Pakistani, and European descent live in the interior and there are some Arabs along the coast [58].

Kenya's Growth Domestic Product (GDP) growth rate is estimated at 4.9% in 2013 and it is estimated at current market prices at US$ 44850 million and with a rate of US$1011 per capita [40]. The economy of Kenya is market-based, with some state-owned infrastructure enterprises, and maintains a liberalized external trade system. The economy's heavy dependence on rain-fed agriculture and the tourism sector leaves it vulnerable to cycles of boom and bust [58]. The proportion of the population below the poverty line is 19.7% in 2005 and the Gini's index is 0.4% in 2009 [43]. However, according to IFAD [59], while Kenya is on the path to economic growth, poverty reduction remains a challenge. More than three quarters of the population lives in rural areas, and rural households rely on agriculture for most of their income. About 70% of the poor are in the central and western regions, living in areas that have medium to high potential to agriculture. Poverty and food insecurity are acute in arid and semi-arid lands which have been severely affected by recurrent droughts. Rural poverty in Kenya is also strongly linked to environmental concerns, especially poor water management, soil erosion, declining soil fertility, and land degradation [59].

According to table 7, Kenya's Human Development Index (HDI) value for 2013 is 0.535, with a rank of 147 out of 187 countries and territories. The table shows also HDI trends since 1980.

Year	Life expectancy at birth	Expected years of schooling	HDI value
1980	57.7	9.3	0.424
1990	59.3	9.3	0.463
2000	52.3	8.3	0.447
2010	56.6	11.1	0.511
2013	61.7 (+4.0)	11.0 (+1.7)	0.535 (+20.0%)

Source: Human development Report, UNDP, 2014 [43]

Table 7. Kenya's HDI trends 1980-2013

Kenya's HDI is above the average of 0.493 of the countries in the low human development group and of 0.502 for countries in Sub-Saharan Africa [43].

4.4.5. Somalia

Somalia is a country located in the Horn of Africa, bordered by Ethiopia to the west, Djibouti to the northwest, the Gulf of Aden to the north, the Indian Ocean to the east, and Kenya to the southwest. It has the longest coastline on the continent's main lands, and its terrain consists mainly of plateaus, plains and highlands. Climatically hot conditions prevail year-round, with periodic monsoon winds and irregular rainfall [60].

The population of Somalia is estimated at 10496 persons, out of which about 47.2% aged 14 years and below, 49.9% aged 15-64 years, while the rest comprise the elderly group (65+) [43]. The majority of the population is ethnic Somalis who have historically inhabited the northern part of the country. Other minorities inhabited the southern part of it [60]. The Somalis are a culturally, linguistically and religiously homogenous people who are divided along clan lines and sparsely scattered over a harsh dry land. The majority of Somalis trace their genealogical origin to the mythical founding father, Samaale or Samaal. However, genealogy constitutes the heart of the Somali social system [61].

A report by UNDP [62] illustrated that Somalia is at a crossroads due to this complex and protracted conflict. The south central region has experienced years of fighting and lawlessness, while the north-east and north-west have achieved a fragile semblance of peace and stability [62]. According to the African Development Bank country brief on Somalia (2013), the country, after more than two decades of civil conflict, is at a turning point in terms of positive political and security developments as well as the commitment of the international community for a sustainable resolution of the protracted crisis in the country. The outlook is more positive than it has been in a generation. However, the situation remains extremely fragile, joint and concerted domestic and international efforts need to be sustained and enhanced to ensure national reconciliation, durable peace, stability and state building [63]. A powerful new vision for Somalia is required, one oriented around building an inclusive peace society, where all people feel empowered, and have the capabilities and opportunities to improve their lives [62].

Somalia is still characterized by a severe lack of basic economic and social statistics. The situation has been worsened by two decades of conflict and the resulting collapse of the country's institutions. The common feature in the structure of the economy of the three sub-entities of Somalia is the predominance of agriculture and livestock as means of livelihood. As well the economy is dominated by the informal sector, in that the economy is based on international trade networks controlled by small groups of wealthy businessmen [63]. Somalia has suffered severe consequences from conflict, as reflected in the indices developed by the Global Human Development Report [62]. In 2011, two consecutive years of failed rainy seasons in the Horn of Africa and the absence of drought mitigation mechanisms resulted in the worst famine and vast humanitarian crisis in the last sixty years in Somalia and the Horn of Africa [63]. This famine signifies an increasingly dismal future, if approaches to both conflict and development do not significantly change. Somalia's Human Development Index (HDI) value is strikingly low at 0.285, which will be further worse, if one accounts for the level of inequality in the distribution of income, education and health. Gender inequality is alarmingly high at 0.776 out of a value of one (complete inequality). In terms of measuring deprivation related to poverty, Somalia's Multidimensional Poverty Index (MPI) of 0.47 out of one would place it at 94 out of 104 countries in 2010. An estimated 82% of Somalis are considered poor across multiple dimensions. The divide between rural and urban population is significant – 61% and 94% respectively. In south central Somalia, 89% of people are poor, across several dimensions, compared to 75% in Puntland, and 72% in Somaliland [62].

4.4.6. South Sudan

South Sudan is a country in northeastern Africa that gained its independence from Sudan in 2011. It is bordered by Sudan to the north, Ethiopia to the east, Kenya to the southeast, Uganda to the south, the Democratic Republic of Congo to the southwest, and the Central African

http://dx.doi.org/10.5772/60012

Republic to the west. South Sudan lies between latitudes 3 and 13 degrees north and longitudes 24 and 36 degrees east. It is covered in tropical forests, swamps, and grasslands [64].

Population in South Sudan, is 11,300,000 persons, and has area of 640,000 km^2 [40]. Of the total population, 15.9% represent those aged 14 and below, 3.5% those who aged 65 and above and the rest comprise the age group of 15 – 64 years. Also about 82.6% of the total population lives in rural areas [43]. South Sudan is historically divided into three provinces: Great Upper Nile, Bahr el Ghazal and Equatoria. For administrative purposes the country is separated into ten states which are then further broken down into 86 counties. The majority of the population continues to live in rural areas, although the urban population of the state capital is steadily increasing. The country's history is the history of competition for natural resources and for influence among different tribes, religions, political factions and colonial powers [65]. South Sudan is potentially rich in natural resources; however, its development was neglected during the colonial period. On the eve of the independence of the whole former Sudan in 1956, the first war broke out. It ended with Addis Ababa Agreement in 1972. Failure of socio-economic development in the post-first war period contributed to the outbreak of the second war in 1983 [66].

4.4.7. *Sudan*

Sudan, the third largest country in Africa, has an area of 1,886,068 km^2 (181 million hectare). It is bordered by Egypt to north, the Red Sea, Eritrea and Ethiopia to the east, South Sudan to the south, the Central African Republic to the southwest, Chad to the west and Libya to the northwest. The Nile River divides the country into eastern and western halves. Along this Nile numerous ancient civilizations were evolved [67].

Sudan geography, like its history, appears to be dominated by the Nile. Most of the population lives along the river, the major cities, industry, wealth and power are all concentrated there. Greater Khartoum, at the junction of Blue and White Niles – comprising the three cities of Khartoum, Khartoum North and Omdurman is overwhelming the larger urban centre in the country. Sudan geography is the product of political and historical factors, as well as, conditions imposed by nature [68]. Population in Sudan, is 36164, about 54.6% aged 14 years and less, 42.0% aged in a range of 15-64 years, while the rest represents 65+ years group. Of the total population, about 65.2% live in rural areas [40].

According to the World Bank report on Sudan [69] Sudan holds the potential to be a regional economic powerhouse. The largest economy in the greater eastern Africa region, it has abundant fertile land and livestock, a reasonable manufacturing base, and strategic market location at the crossroads of sub-Saharan Africa and the Middle East. Oil discovery and export in the last decade fueled unprecedented growth (GDP grew more than six-fold from 1999 to 2010). However, much of this potential has not been realized due to long-running conflict and governance challenges; and the gains of the last growth decade have not advanced productive public investment that contributes to poverty reduction, or been widely shared [70]. During 1970-90 the real GDP growth rate fell below -5% in six years (1972, 1978, 1979, 1984, 1985, and 1990) and exceeded 10% in four years (1974, 1975, 1976, and 1987). In contrast, the growth rate has hovered within 5-11% range during 1999-2008 periods [74]. Currently, Sudan's Growth

Domestic Product (GDP) growth rate is estimated at 3.6% in 2013 and it is estimated at current market prices at US$ 70,463,000 and with a rate of US$1,856 per capita [40].

According to table 8, Sudan's Human Development Index (HDI) value for 2013 is 0.473, with a rank of 166 out of 187 countries and territories. The table shows also HDI trends since 1980.

Year	Life expectancy at birth	Expected years of schooling	HDI value
1980	54.2	3.7	0.331
1990	55.5	3.9	0.342
2000	58.0	4.5	0.385
2010	61.5	7.3	0.463
2013	62.1 (+7.9)	7.3 (+3.6)	0.473 (+42.8%)

Source: Human development Report, UNDP, 2014 [43]

Table 8. Sudan's HDI trends 1980-2013

Sudan's HDI is below the average of 0.493 of the countries in the low human development group and of 0.502 for countries in Sub-Saharan Africa [43].

According to the World Bank report (2013) gender disparities remain persistent in Sudan. Women comprise only23% of the formal economy, but 70% of the informal economy, with a majority of them engaged in agricultural production. On the other hand, despite those ten years of oil boom, Sudan continues to suffer wide and deep swaths of poverty and stark inequality between regions [69].

4.4.8. Uganda

Uganda is a landlocked country in east Africa. It is bordered to the east by Kenya, to the north by South Sudan, to the west by the Republic Democratic of the Congo, to the southwest by Rwanda, and to the south by Tanzania. The southern part of the country includes a substantial portion of Lake Victoria, shared with Kenya and Tanzania, situating the country in the African Great Lakes region. Uganda also lies within the Nile basin, and has a varied but generally equatorial climate. Uganda takes its name from Buganda Kingdom, and beginning in the late 1800s, the area was ruled by the British, who established administrative law across the territory. Uganda gained independence in 1962. The period since then, has been marked by intermittent conflicts, most recently a lengthy civil war against the Lord' Resistance Army, which has caused tens of thousands of causalities and displaced more than million people [71].

Population in Uganda, is 35357, about 50.9% aged 14 years and less, 47.7% aged in a range of 15-64 years, while the rest (1.4%), represents 65+ years group. Of the total population, about 81.9% live in rural areas. Uganda, in 2013 had a GDP of US$ 23,459,000 and a per capita income of US$ 624,000 and the GDP growth was at 5.2% rate [40].

According to table 9, Uganda's Human Development Index (HDI) value for 2013 is 0.484, with a rank of 164 out of 187 countries and territories. The table shows also HDI trends since 1980.

Year	Life expectancy at birth	Expected years of schooling	HDI value
1980	49.5	4.0	0.293
1990	47.5	5.7	0.310
2000	48.1	10.7	0.392
2010	57.3	10.8	0.472
2013	59.2 (+9.7)	10.8 (+6.8)	0.484 (+65.0%)

Source: Human development Report, UNDP, 2014 [43]

Table 9. Uganda's HDI trends 1980-2013

Uganda's HDI is below the average of 0.493 of the countries in the low human development group and of 0.502 for countries in Sub-Saharan Africa [43].

Uganda's sustained growth over the past two decades has continued to be rewarded with strong poverty reduction. However, poverty remains mainly a rural phenomenon, with 34% of the rural population living the national poverty line. The poorest areas of the country are in the north, where poverty incidence is consistently above 40% and in many districts exceeds 60% [72].

5. Farming systems in IGAD region

5.1. Introduction

According to Giessen [13] the hydrological assets in the Horn of Africa includes precipitation, the Nile River and its tributaries, the Ethiopian Highlands which make the water tower of the region, and a number of some important lakes such as Lake Tana, Lake Turkana and Lake Victoria. The Horn of Africa is not a total desert or dry wetland, as it is often considered. Vegetation is quite varied and spatially correlated with precipitation levels and the systems of rivers and lakes. Very generally, the drier eastern parts of the Horn of Africa consist of desert, semi-desert and steppe vegetation, while in areas close to the major hydrological assets savannah grasslands and deciduous forest vegetation occur. In terms of biodiversity, Ethiopian highlands can be regarded as historical gene centre or a centre for origin for many crops. As well high value protected areas are also found in Kenya, South Sudan and Uganda [13].

In IGAD countries, agriculture dominates their economies in terms of contribution to GDP, employment and income. This indicates clearly that, for sound socio-economic development, the real growth and developing of this sector will be the leading engine. Any developments in this sector are vital for poverty reduction, wealth creation and improved food security status.

Most of the landscapes in the IGAD region consist mainly of lowlands with arid, semi-arid or dry sub-humid zones. Based on agro-ecological zoning considerations, the region can be classified as arid (55%), semi-arid (15%), sub-humid (16%) and humid (2%) and high lands zones (12%). About 3.4 million km^2, or 80% of the IGAD region's total area, consists of lowland, with arid, semi-arid or dry sub-humid climates, where precipitation is low and uncertain (100 – 600 mm per annum) [73]. These arid and semi-arid lands, according to Mortimmore, (2009) provide numerous goods and services that have great economic, social, cultural and biological value in all geographical aspects: locally, nationally and globally [74]. More than 40% of the total area is unproductive because of severe environmental degradation, resulting from both natural conditions and human actions. The agricultural sector is one of the three priority development areas of IGAD and sector oriented activities are focused on food security, natural resources management and environmental protection. The majority of the people in the region depend on natural resources for their livelihood. The different key components of drylands (land, water, nutrients, and energy) are deeply interconnected: changes in one component will affect the other [75].

According to Knips, [76] IGAD region could be divided into seven ecological Zones, namely: desert, arid, semi-arid, sub-humid, humid, highland and urban/peri-urban ecological zones (Figure 2).

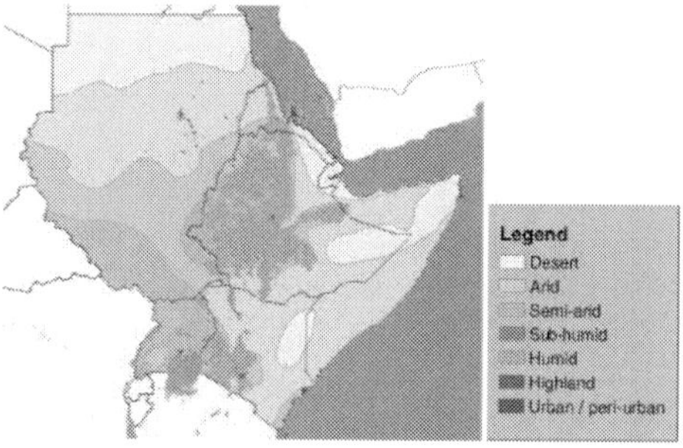

Figure 2. A map showing the agro-ecological zones in IGAD region

5.2. Major farming systems in the region

As well documented in the literature, and based on Schiere, [14] farming system typologies are dictated by climate, production goals and culture with a farming system being described as a unit consisting of a human group (usually a household) and the resources it manages in its environment, involving the direct production of plant and/or animal products [14].

Generally, it has been reported by Robinson *et al.* (2011) that, the existing global system classifications were facing the following limitations: (i) they tend to focus either on crops or on livestock farming; (ii) some classification systems tend to group the majority of production systems into a single mixed farming category, which may include many different combinations of crop and livestock species; (iii) many existing classification systems can be useful at very broad scales but they are often of little practical use for priority setting and planning at national level [28].

http://dx.doi.org/10.5772/60012

The countries of the Horn are characterized by four broad-based systems of land utilization. These systems are pastoralism, agro-pastoralism, rain-fed and irrigated agriculture. However, it has to be emphasized that these four systems are closely inter-linked through symbiotic relationships. High potential areas which are put under agriculture have a better chance of supporting the system that utilizes them. Yet the increasing demand for food production has put marginal lands in the region under severe strain and has led to long lasting land degradation [77]. Pastoralists and agro-pastoralists represent a high proportion of the population in the region and are utilizing the arid and semi-arid areas. This is largely due to the fact that much of the land in the Horn is dry land which offers little or no opportunity for means of subsistence other than livestock raising [78]. According to Sandford, [79] in the late 1970s the most important single countries of the world in terms of the numbers of pastoralists were Sudan, the USA, Somalia, Chad, Ethiopia, Kenya, Mali, Mauritania, India and China [79]. The order may be different at present, but one important thing to note is that four out of the ten countries above are found in the Horn of Africa, which indicates the importance of pastoralism as a livelihood system and as a method of land utilization in the region [78].

In African context, Dixon *et al.* [16] defined a farming system as a population of farm households, often a mix of small and large farms, that as a group have broadly similar patterns of livelihood and consumption patterns and constraints and opportunities, and for which similar development strategies and interventions would be appropriate. Often such systems share similar agro-ecological and market access conditions. Based on this definition and on two factors, namely the available natural resource base and the dominant patterns of consumption and household livelihoods, they mentioned about 13 farming systems in Africa in general; these systems are:

1. Maize mixed farming systems: sub-humid and humid, in east, central and southern Africa.
2. Agro-pastoral farming systems: semi-arid areas, west, east and southern Africa.
3. Cereal root crop mixed farming system: sub-humid areas, west and central Africa.
4. Root and tuber crop farming systems: lowland areas, west and central Africa.
5. Highland perennial farming systems: moist highland areas, east Africa.
6. Highland mixed farming systems: cool highland areas, east and southern Africa.
7. Humid lowland tree crop farming systems: humid lowland areas, west and central Africa.
8. Pastoral farming systems: in arid areas, west, east and southern Africa.
9. Fish-based farming systems: along coastal areas and major lakes.
10. Forest-based farming systems: humid lowlands, central Africa.
11. Irrigated farming systems: low rainfall areas.
12. Sparse arid pastoralism and oases farming systems: arid areas, west, north east and southern Africa.
13. Urban and peri-urban farming systems: all parts of Africa

Based on this classification, the following farming systems are currently practiced in IGAD region:

1. Maize mixed farming systems: this is found in Kenya
2. Agro-pastoral farming systems: such as in Somalia, Djibouti and Ethiopia
3. Highland perennial farming systems: this one found in Ethiopia and Uganda
4. Highland mixed farming systems
5. Pastoral farming systems: in the arid and semi-arid zones of Sudan, Eritrea, Ethiopia, Somalia and Kenya and Uganda.
6. Fish-based farming systems: such as in Kenya.
7. Irrigated farming systems: this comprises large scale irrigation schemes such as the Gezira scheme in Sudan and Wabi Shebelle in Somalia
8. Sparse arid pastoralism and oases farming systems: this is found in Sudan
9. Urban and peri-urban farming systems: around the major cities in all IGAD countries.

According to a regional study on the livestock sector in the horn of Africa [80] commissioned by the African Development Bank, there are two main production systems practiced in the Horn of Africa accounting for over 80% of the total livestock numbers. These are the pastoral and agro-pastoral production systems; however, there is also the mixed crop-livestock production system [80]. On the other hand, following an approach tackled by FAO, the Africa Development Bank study categorized the production systems in the Horn of Africa region into seven different production systems. These systems, according to AfDB study [80] could be summarized as follow:

1. Grassland based systems: these systems in the Horn of Africa could be sub-divided into three sub-types:
    a. Livestock only, arid/semi-arid tropics and sub-tropics production system: it has a growing period of less than 180 days per year; the main livestock type is grazing ruminants. It is mainly pastoral due to scarcity of rainfall, water and pasture. One of the most severe problems of this system is the feed variability, in addition to some environmental concerns and problems of land degradation.
    b. Humid and sub-humid tropics and sub-tropics production system: this enjoys more than 180 days of pasture growing period. It is found in South Sudan, Ethiopia and Uganda. Among the main constraints are the prevalence of trypanosomiasis, poor feed quality, poor road infrastructure and some environmental concerns.
    c. Temperate zones and tropical highlands production system: it is practiced in Ethiopia, South Sudan, Uganda and Kenya. It has seen the introduction of temperate livestock breeds with some efforts of local breeds improvement through artificial insemination. Range is the primary feed source making the livestock vulnerable to changing weather patterns. There is potential for greater forage production similar

to the intensive systems in other regions, however, the balance between intensive production and ecosystem protection remains the bigger challenge for increased productivity.

2. Mixed irrigated husbandry production system: this system is only reported in Somalia under the type mixed irrigated arid and semi-arid in which irrigation makes year round intensive crop production feasible. In other countries it is thinly spread with efforts going on in Ethiopia and Sudan, while Kenya has put in place policies that are geared at promoting this system in the drylands. Private investments in parts of northern Somalia has made some production possible under this system.

3. Mixed rain-fed systems: these are found in all the countries of the Horn except in Djibouti and Eritrea. There are three main categories of this system in the Horn of Africa:

   a. Temperate zones and tropical highlands production system: it is practiced in the tropical highlands of Ethiopia where the large numbers of livestock provide a variety of services in support of crop production, Kenya where dairy development and dairy cattle improvement has taken root and a little bit in Sudan and Uganda. It has been the most versatile in response to technological innovation, adaptation to climate change and changing land use practices. The future challenge of this system is the adoption of cleaner production practices that conserve the environment while at the same time coping with the increasing demand for livestock products.

   b. Humid and sub-humid tropics and sub-tropics: it is based on a mixed farming method under varying socio-economic and climatic conditions. It is found in Uganda, Ethiopia and Kenya. The challenge for this system is to device ways of increasing productivity under reduced land resource availability.

   c. Arid and semi-arid tropics and sub-tropics production system: is a mixed production system with a vegetation growth period of less than 180 days. The soils are normally less productive and rainfall is usually too low to sustain cropping. Livestock is normally the primary income generating activity. Crop production is very low and normally for subsistence only. Rangeland degradation and high methane production per animal are among the challenges of this system.

Within the agricultural sector in IGAD region, the major contribution to the GDP comes from livestock. The importance of the livestock sector in the region can partly be explained by the fact that the major proportion of the land area in the region is classified as arid, with highly variable rainfall making it unsuitable for crop production. This leaves livestock production as the only viable form of land use. In agro-ecological zones where crop production is possible it is mostly practiced in mixed systems with livestock providing important inputs into the farming system. Livestock production systems in the region are pastoral, agro-pastoral, settled mixed crop-livestock production systems and small-scale dairy production [73]. The pastoral production systems according to Otte and Chilonda, [81], are characterized by a contribution of livestock and livestock-related activities to household gross revenue of over 50% and it involves seasonal or annual mobility of livestock in search of pasture over a large area of rangeland [81]. Both pastoral and agro-pastoral livestock systems are practiced in all IGAD countries. The dominant species of livestock kept under both systems are cattle, sheep, goats and camels. The degree of mobility of herds and households varies depending on environmental factors and normally increases with the increase of aridity [73]. In the agro-pastoral systems, the livestock are kept for subsistence (milk and milk products), transportation (camels, donkeys), land preparation (oxen, camels), sale or exchange, saving, bride wealth and insurance against crop failure. The population generally lives in permanent villages, although part of the herds may continue to migrate seasonally in the care of herd boys. The main crops planted in this system are millet, sorghum, maize, and cowpea. Irrigation is rarely practiced, except for few locations in Somalia and Sudan, where cotton, sugarcane and rice are grown [16].

Settled mixed systems are found in the highlands, sub-humid and humid zones within the IGAD countries. They are predominantly subsistence oriented and crop dominated with the type of crops planted depend on agro-climatic conditions, while the numbers and species of livestock kept depend on human population pressure and prevalence of trypanosomosis. It is found in all IGAD countries except in Djibouti and Somalia [73]. The major characteristic of the small-scale dairy production systems is the production of milk for sale. Milk production is integrated with the growing of subsistence crops such as maize, beans and potatoes and of cash crops such as coffee, tea and pyrethrum. This system is found in Kenya, Ethiopia and Uganda [81].

Globally and at regional levels, in the drylands context, the agro-ecosystems comprise a diverse and complex mix of pastoral, agro-pastoral, rain-fed and irrigated farming practices. Farmers and pastoralists employ a diverse mix of food, fodder and fiber crops, vegetables, rangeland and pasture species, fruit and fuel wood trees, medicinal plants, livestock and fish to meet their food and livelihood needs. They have developed these practices over centuries, adapting them to the limited resources and variable climate that characterize dry areas. Agricultural production systems in the drylands face not only persistence water scarcity and frequent drought, but also high climatic variability, land degradation, desertification and widespread poverty. These constraints are expected to intensify as a result of climate change [82]. These areas are usually perceived as having low production potential, but according to Fowler and Hodgkin, [83] they are home to several important centers of origin and diversity of crops, vegetables, livestock, trees and fish, and most traditional farming systems maintain this agro-biodiversity. These genetic resources can provide breeders with the traits needed to adapt plants and animals to heterogeneous and changing environments [83]. These genetic resources, as reported by Maestre *et al.* [84] are an important buffer against the effects of climate change and desertification [84]. However, as mentioned earlier, and according to Harvey et al., (2011) land degradation and pressure on natural habitats threaten biodiversity in dry areas and farmer behavior, if not directed otherwise, will generally result in decline in species diversity to meet immediate production objectives[85]. Therefore, the multifaceted constraints facing dryland agricultural systems call for broad-based, integrated approaches addressing the full range of socio-economic and biophysical constraints that farmers and pastoralists in drylands typically face. This requires innovative approaches that bring together all participants in the impact pathway, from primary producers to policy makers, to develop technologies, resource management strategies, and institutional arrangements that: reduce demand for water per unit crop area, and livestock unit; improve water capture and storage; increase productivity per unit of water and land at farm and landscape scales; enhance the capacity of communities and the most marginalized actors within them; and strengthen institutional arrangements to build resilience of livelihoods and increase system productivity through diversification and sustainable intensification [82]. These approaches aim to identify, quantify and address the driving forces and interactions that shape and constrain farming systems and the management of natural resources [86].

To give full overview about the major farming systems currently practiced at country level in the region, a detailed description and analysis will be presented on the following part of this chapter.

5.3. Farming systems in Djibouti

Geographically, Djibouti has one of the most inhospitable, barren environments on the planet. It has virtually no arable land, no permanent fresh water source, no significant mineral resources, very little vegetation, high daily temperature and severe persistent drought for the past six years [50].

According to the Convention of Biological Diversity [87], Djibouti is made up of rich and varied terrestrial and aquatic eco-systems. The majority of the country is defined as desert with a climate that is torrid and dry throughout the country. It has three distinct geographic areas: the coastal plains, the volcanic plateaus in the southern and central parts and in the north, the mountain ranges where the elevation can be as high as 2000 metres above sea level. Djibouti contains several different types of eco-systems however over 90% of the land is desert. The terrestrial eco-system is separated into the mountains region and the semi-arid lowlands which are dominated by shrubs and trees [87]. Almost 78% of the people in Djibouti base their livelihoods on agriculture. Therefore, most parts of Djibouti are allocated for agricultural activities, all the land is used for pasture, with smaller areas for the production of crops, mainly vegetables and fruits [13]. The National Strategy for Food and Nutrition Security (2008) estimated that less than 10% of the calories consumed nationally come from domestic production, the remaining 90% being imported from neighboring countries or from the international market. This heavy reliance on food import makes the country highly vulnerable to external market risks that are often beyond its control [88].

According to FAO [89] livestock rearing is the main livelihoods activity for 80% of rural households. In recent years, herds have been decimated due to impact of recurrent drought particularly that of 2010/2011 [89]. In rural areas, two traditional production systems are present: pastoralism and small scale farming. Pastoralism is an age-old and deeply entrenched tradition that dominates Djibouti's rural economy. Pastoral activities consist primarily of extensive nomadic herding which often represents the sole source of subsistence for pastoral communities. Some 90.5% of the country's territory can be classified as pastoral lands that are used for herding. Transhumance is still practiced extensively along grazing routes determined by the presence of water and pasture. Mobility is a highly efficient way of managing the sparse vegetation and relatively low fertility of fragile soils of Djibouti [88].

From the previous discussion and evidences from the literature pertaining to the currently practiced farming systems in Djibouti are crop farming, pastoral and agro-pastoral farming systems. According to Brass, [89] crop farming has been introduced to Djibouti in the past 30 years, and comprises only a tiny percentage of Djibouti's economy, population and land. The remainder of Djibouti's non-urban lands is used solely for livestock production, the vast majority of which is subsistence nomadic or semi-nomadic pastoralism of small ruminants (primarily goats) and camels. The Djiboutian livestock production system is unifies with its neighbours; pastoralists move between Djibouti, Ethiopia and Somalia following the rains. These pastoralists maintains a traditional approach to animals, seeing them as store of value and a source of protein via milk – animals are only slaughter or sold in case of liquidity crisis or for important celebrations. Thus, while it is the prominent rural activity, livestock production is not well integrated into the national monetary economy, nor t is a government priority [90].

5.4. Farming systems in Eritrea

Eritrea's topography can be divided into three broad categories: the arid, narrow, lowland along the Red Sea; the north central region, which is an extension of the Ethiopian plateau and is dissected by river valleys; and the western plain along the Sudanese border. The highest point is Emba Soira, southeast of Asmara at 3,010 metres; the lowest is in the Dinakil Depression along the Red Sea, which at places is at 130 metres below sea level. It is one of the hottest places on earth. The Setit River is Eritrea's only perennial waterway. It flows from Ethiopia, where it is called Tekeze, along the western border and into Sudan, where it is called the Atbara. There are other seasonal rivers during the rainy season, namely the Anseba, the Baraka, and the Mareb (the Gash) [91]. Eritrea is geographically situated on the south-eastern border of the Sahel-Zone. Rainfall ranges from 1200 to less than 200 mm per year. The interregional differences and the variability in amount are high, and the risk to have an insufficient rainy season with erratic rainfalls is high. Some parts have hot tropical semiarid climate, with rainfall regimes that range from 400 to less than 700 mm per year [92]. As one of the arid or semi-arid Sahelian countries of Africa, Eritrea faced serious droughts in 1975, 1984,1989, and 1991 Less than 5% of the land in Eritrea is arable, and permanent crops occupy a mere 0.03% of total land area. Most land is suitable for pasturage, but some areas, such as the Red Sea coast and the far north, are too arid even for this purpose [91]. However, land in Eritrea falls into four categories, namely cultivated, grazing, forests and barren land [93]. More than 70% of the population depends on traditional subsistence agriculture for their livelihood. The main sources of income for rural households are the sale of crops, livestock, and livestock products; wages for daily labour and remittances. In urban areas, people generate income from wage labour, small businesses, petty trade, and poultry farming. Eritrea has a number of agricultural systems: rain-fed cereal and pulses, semi-commercial and peri-urban agriculture, small-scale irrigated horticulture, commercial farming, agro-pastoral rain-fed farming, and agro-pastoral spate irrigation systems [93]. The agricultural sector is hampered by the absence of modern farming equipments and techniques, erratic rainfall, exhausted soils and lack of financial services and investment. Major agricultural products are barley, beans, lintels, dairy products, meat, milk, skins, sorghum, teff and wheat [91]. Eritrea's location in arid and semi-arid zones makes it vulnerable to the adverse effects of climate change, such as drought, pest infestation and degradation of natural resources which can affect food security, and if it does, can eventually lead to malnutrition or under-nutrition in adult and children [93]. Despite the fact the economy is agriculture based; its contribution to the GDP has been moderate, due to recurrent drought, rudimental farming methods, and effects of war. Now, this sector could be transformed into a promising sector through increased reliance on irrigation and improved farming methods and promotion of the livestock sector exports activities especially to the Middle East Markets. There are also opportunities in the production of high valued crops and vegetables for exports to Europe and Middle East [94].

5.5. Farming systems in Ethiopia

With its dramatic geological history and broad latitudinal and altitudinal ranges, Ethiopia encompasses an extraordinary number of the world's broad ecological zones. With a high plateau and a central mountain range divided by the Great Rift Valley, Ethiopia contains a huge altitudinal range from the depressions in the Afar (115 metres below sea level) to the mountain tops of Ras Dashen in the north (4,533 metres above sea level) and the Bale Mountains in the southeast. The headwaters of the Blue Nile are located in northwest Ethiopia at Tana Lake. This range of habitats also supports a rich variety of species, which contributes to the overall biological diversity of the country [95].

The agricultural sector greatly influences the rate of economic growth in Ethiopia: about 11.7 million smallholder farmers account for approximately 95% of agricultural GDP and 85% of the population. With a total area of about 1.13 million km^2 and about 51.3 million hectares of arable land, Ethiopia has tremendous potential for agricultural development. However, only about 10.6 million hectares of land are currently being cultivated, just over 20% of the total arable area. Ethiopian agriculture is dominated by a subsistence, low input, low output, and rain-fed farming system. Low productivity levels could be attributed to limited access by smallholder farmers to financial services, improved production technologies, irrigation and agricultural markets; and more importantly to poor land management practices that have led to severe land degradation [96].

According to Rahmato, [97] Ethiopia has varied agro-climatic zones. The Government extension programmes list these as: areas of adequate rainfall; areas of moisture stress; and pastoral areas. Farmers traditionally classify them as dega (cool), woina dega (temperate) and qolla (low land, warm climate). This diversity makes it a region for growing a variety of crops [97]. The country is endowed with one of the most bio-diverse systems in the world. It has earned the name 'the Water Tower of Eastern Africa' for having more than ten rivers, each of which has irrigation potential [98].

Based on John Dixon et al, [16] and FAO report [99], the following farming systems are found in Ethiopia:

1. Irrigated, small scale, traditional farming system: these have been established under self-help programs and initiatives by farmers' groups with sizes varying from a few to 100 ha. Modern small scale irrigation is also practiced. Traditional irrigation is practiced in the different agro-ecological zones making use of rivers, creeks, or gully waters that can last

up to three months in the dry season. These systems are less capital-intensive than large scale ones and are managed by traditional community rules and water rights, which make them an integral part of indigenous farming systems. Irrigated agriculture in Ethiopia is not well developed, but includes both traditional and modern small scale systems. The typical cropping pattern is organized into dry and wet seasons. In the dry one (September-April) vegetables like potato, onion and peppers. In the wet season (May-November) all cereals including teff, barley and rice are grown. Commercial agriculture is a relatively recent introduction and its contribution to total agricultural production is therefore still small.

2. Pastoral farming systems: this lies where climatic conditions lead to sparse vegetation cover over often fragile soils, with a scarcity of surface water. It can support only sparse human population and is not suitable for rain-fed crop production, only well adapted livestock. It is governed by social and community laws concerning the management and utilization of natural resources. Although there are profound similarities, the different ethnic groups practicing the system exhibit considerable differences in their overall enterprise patterns, seasonal movements, and the other natural resource-based activities. There are also wide variations in the level of integration of pastoralists into the market economy. Typical pastoral family size is relatively large and has a high dependency ratio, mainly due to cultural factors. It is common for men to have more than one wife, which is believed to make mobility of livestock from place to place easier. The nature of animal husbandry and the mobility of herds demand extended families in order to distribute the workload and defend economic and/or social/cultural interests. Better off pastoral families create employment for poor and marginalized families. Livestock production is dependent upon climate, vegetation, and animal type. The main sources of feed in pastoral areas are grasses, shrubs and browse. Their distinctive management systems, includes undertaking seasonal migrations in response to availability of grazing; and keeping different species of livestock, including small and large ruminants. Cattle and sheep are kept in areas with reasonably abundant water and where grazing species are predominant; goats and camels are reared in drier areas, where browse species predominates. This farming system links up with the rest of the agricultural economy in that male cattle are sold to highland farmers for draft power. There is also linkage in the opposite direction as pastoralists import breeding bulls from the highlands to bring new blood into their herds.

3. Agro-pastoral farming system: this is found under conditions fairly similar to those of the pastoral system, the key difference being the slightly greater availability of water. It is a semi-nomadic livelihood in which livestock production is dominant. Crops are grown but play a less significant role than in most other farming systems as moisture stress is a critical limiting factor on crop production. It extends from the north-western to south-western parts of the Nile Basin. It is practiced along river banks in particular and in areas neighboring the pasture zones which receive slightly higher rainfall. The average family owns between 1.0 to 1.5 ha of cultivated land, a minimum of 6 cattle and more than 6 goats. Main crops are sesame, cotton, sorghum and vegetables. Additional cash is obtained from sale of fruit, vegetables, honey, gum, firewood and grasses. Agro-pastoralists benefit from

being able to graze their animals along the perennial Nile tributaries, which help them cope in drought periods. The major challenges to this system include ecological disturbance and land degradation associated with slash and burn shifting cultivation, moisture stress and lack of adequate water supply, human and livestock diseases, livestock feed shortages during the dry season and poor infrastructure. Major constraints include sociological factors, transport and communication, marketing, livestock diseases, land tenure, livestock diseases, marketing, inter-clan disputes, low rainfall, lack of security in border areas and disputes between agro-pastoralists and settled farmers.

4. Dryland farming system: It is undertaken in rangelands ecosystems where there is sufficient soil moisture or ground water to allow settled farming. It has many similarities with the agro-pastoral system. The main difference being in the relative importance of the arable and livestock components, where in this system, crops are more important. It is rainfed system, where mixed crops are grown such as sorghum, teff, wheat, maize, millet, sesame, groundnut and some vegetables. Livestock in this system is indigenous and their management is traditional. The constraints of this farming system include: short and long term droughts, low productivity, use of local crop varieties and landraces, lack of inputs, and pests and disease; deforestation, insufficient fodder and forage, animal diseases, lack of saving and credit institutions, and marketing bottlenecks.

5. Highland temperate farming system: this is found extensively in the high land complexes comprising mountain chains and plateaus at altitudes of 2000 to 3000 m above sea level. Traditional rainfed subsistence farming is practiced on the highland plateaus. Teff, wheat, barley, sorghum, broad beans, field peas, chickpeas, vetch and oil crops are grown. The livestock in this system include cattle, sheep, goats, horses, donkeys, mules, poultry and bees. The major sources for feed of livestock are natural pastures with contribution from crop residues, industrial by-products and locally grown fodders. There is growing pressure on land due to high rates of human and livestock population growth in the high land areas that had led to severe shortage of land and deterioration of natural resources base. The constraints include weed infestation, crop pests and diseases, use of traditional tools, post-harvest crop losses, various types of widespread livestock diseases, and climate change.

6. Highland cold farming system: is practiced at more than 3000m above the sea level. The climate is cold with frequent night frosts during the dry season. This system features two components, barley cropping and grazing in the lower part (up to 3300 m) and pure grazing above that altitude; most farmers use both components. Livestock are a particularly important component as the crop yields are low and unreliable. The constraints include: land not inherently suitable for cropping, little knowledge of pasture improvement, livestock health problems, poor access to infrastructure, human population pressure and low incomes.

7. Forest based farming system: this is found in southern Ethiopia, with rain falls all the year, with very short dry period from December to February. Traditionally communities depended on extraction of forest products, hunting, gathering and some pastoral livestock herding. Shifting cultivation is practiced to grow maize, sorghum and millet. Vegetables,

fruit trees, spices and coffee trees are also cultivated. The constraints include: destruction and over-exploitation of natural resources, ecological disturbances and emergence of pests and weeds, accelerated loss of soil nutrients, land degradation, animal diseases and poor services to livestock sector.

8. Riverside and lakeside farming system: livelihoods in this system are based on crop cultivation supplemented by fishing and livestock rearing. The system is practiced along the riverbanks and in the swamps around Lake Tana. Crops cultivated include maize, sorghum, sesame, groundnut, cowpea, rice, onions, other vegetables, tobacco and beans.

9. Market oriented agriculture (including urban, peri-urban and commercial farming): it encompasses a wide range of specialized forms of agriculture which developed to cater for urban and export sector. Most of the large scale production farms are concentrated in the western lowlands and the north Gondar and west Gojam areas. Most farming operations are carried out with machinery and make wide use of fertilizers and other yield enhancing inputs. This farming system is very heterogeneous, ranging from small scale, capital intensive, market-oriented, vegetable-growing, dairy farming, and livestock fattening, to part-time farming by the urban poor to cover part of their subsistence requirements. The level of crop-livestock integration is often low. The main shortcoming include: low yields, high cost of production, shortage of credit, poor infrastructure, poor access to marketing, land degradation and uncertain land tenure.

5.6. Farming systems in Kenya

Kenya is ecologically diverse, and most of its land is classified as arid or semi-arid, yet also higher elevations lush montane forests are found. It is the home to the second-highest peak in Africa, glacier-capped Mount Kenya [100]. It has a great variety of agro-ecological conditions broadly correlated with altitude and aspect, ranging from arid pastoral rangelands to tropical alpine conditions [101]. By area, about 28% of Kenya's ecosystems are marine and 72% are terrestrial. Kenya's average annual rainfall is approximately 630 mm per year, but it varies across the country. It ranges from 200-400 mm per year in northern and eastern Kenya to up to 2000 mm per year in the highlands and mountains of the southwest. More than 80% of Kenya is arid and semi-arid. Croplands and the associated agro-ecosystems cover about 19% of Kenya [102]. According to Sambroek et al. [103] Kenya's land is divided into seven agro-ecological zones based on moisture index. These are humid, sub-humid, semi-humid, semi-humid to semi-arid, semi-arid, arid and very arid [103].

The major ecosystems in Kenya are the terrestrial, aquatic, marine and endangered ecosystems. Regarding the terrestrial ecosystems, according to WRI, et al. [104] a mosaic of grassland savanna, wood-land savanna, and bush land plant communities are found in Kenya's ASALs, depending upon soil type, rain-shadow effects, and other factors. In areas with rainfall above 800 mm per year, the potential natural vegetation is forests which cover about 20% of Kenya and along the Indian Ocean; there is a belt of forest comprising about 1.7% of the country [104]. These ASALs lie in agro-climatic zones IV, V and VI and they represent 80% of the country's area and support 25% of the human population and 50% of the total livestock population. According to Barret *et al.* [105] these ASALs are ill-suited for intensive crop production but

http://dx.doi.org/10.5772/60012

fairly suited for extensive livestock production [105].The aquatic ecosystems according to NEMA and UNDP, [106] cover about 8% of Kenya's surface area and include freshwater and saline lakes, rivers and wetlands [106]. Kenya's territorial sea area is approximately 13800 km^2 covered by mangroves at about 600 km^2. This in addition to coral reefs, sea grasses and sandy beaches [102]. In general, the largest proportion of the country falls into two categories of land cover: (i) herbaceous cover, closed open, and (ii) sparse herbaceous or shrub cover. The two types of land cover are ideal for pastoralism (sheep, goats and camels) which characterizes the livelihoods of the arid and semi-arid lands (ASALs) that cover 83% of the country's land area. Cultivated and managed areas are to be found in agriculturally high- to medium-potential areas of the country. Areas with tree cover are found in the major water catchments and along the coastal strip. The country's grain basket, the Rift Valley and Western Provinces, is characterized by mosaic cover: cropland, trees or other natural vegetation. Natural resources from wildlife and forestry in the protected areas play two basic roles in development: support to subsistence livelihoods, and source of earnings from tourism [100].

Based on John Dixon et al, [16] and FAO report [99], the following farming systems are found in Kenya:

1. **Maize mixed farming system:** it is one of the most important food production systems that mostly found at altitudes of 800-1500 metres. It is characterized by farm sizes less than 2 ha, with scattered irrigation schemes. Maize is the staple crop with other cash crops, cattle and small ruminants. Among the challenges faced by this system: input shortages, declining soil fertility, drought, market volatility and incidence of chronic poverty.

2. **Irrigated farming system:** in Kenya, irrigation may be seen as both a major cause of and an important solution to the country's increasing water scarcity and water insecurity. In addition to the traditional small scale irrigation practices, large scale irrigation schemes have also been in existence from the time of the colonial era. Irrigation schemes in Kenya can be categorized into three organizational types: (a) smallholder schemes, these are schemes of variable farm sizes and are operated by water user groups or by farmers' organizations within the scheme. Approximately, at present there are 2,500 such irrigation schemes covering an area of about 47,000 ha which accounts for about 46% of the total irrigated area, with overall management from the Ministry of Water and Irrigation. Their produce is mostly for domestic food consumption and local market. The challenge faced by this type include marketing problems and poor access to credits; (b) large scale schemes managed by National Irrigation Board, these range from several hundred to several thousand hectares in size that produce for domestic and export markets. Today, there are a total of seven such schemes covering an area of 13,000 ha, accounting for about 12% of Kenya's irrigated land and about 12% of the farmers active in irrigated agriculture. About 90% of the Kenya's rice is produced in these schemes. One bottleneck is the financial sustainability of these schemes; (c) Commercial flower and vegetable farms, also known as agro-industrial irrigation of high value crops, these are schemes with modernized irrigation facilities. They produce almost for export markets in an area of 42,800 ha representing about 42% of the land under irrigation in Kenya. They financed and developed by private corporations or individuals and relying heavily on pump-based technol-

ogies in combination with drip or sprinkler irrigation. Their main concern is the maintenance of their international competitiveness. The overall major challenge for the three types is the secure access to water [107]. However, Blank, et al., [108] highlighted that the boundaries between these three types have become blurred due to some rapid changes. Among these changes: the collapse of the government capacity to manage the large scale systems, the commercialization of smallholder irrigation has enabled some individuals to move away from irrigation of traditional crops and enter the high value vegetable and fruit market, and the third change is that new technologies are rapidly being introduced and adopted widely by irrigators [108].

3. **Pastoral farming system:** according to Cecchi et al, [109] this system in Kenya cover around 336,367 km^2, with estimation of rural population of about 2,048,000. The main livestock species raised in this system are camels, cattle, sheep and goats [109]. Among the main constraints of this system are the recurrent drought, insecurity, water accessibility, rising poverty, declining asset level, environmental degradation and desertification [110].

4. **Agro-pastoral farming system:** is a form of livelihood that combines crop production with pastoralism. In Kenya, the area covered by this system estimated at about 112,081,000 km^2, with rural population of around 6,392,000. The main livestock species raised in this system, cattle, goats, sheep, camels, donkeys, poultry and pigs [109].

5. **Urban/pre-urban farming system:** as in other countries of Sub-Saharan Countries, farming in Kenya is very common among urban households, particularly poor female-headed ones. In Nairobi for example, four sub-types of this farming system are found, namely; small scale subsistence crop production, small-scale market oriented crop production, small-scale livestock production and large-scale commercial farming.

5.7. Farming systems in Somalia

Somalia is an arid region in the north and semi-arid in the south. The climate of seasonal rains, almost continuous winds and even worse, irregularity of rainfall over the years, made it very difficult for the population to work effectively. In areas where rainfall is sufficient for cultivation, there was the problem of shortage of agricultural workers. Animal husbandry was always limited by the scarcity of water. About four-fifths of the population of Somalia was engaged in agriculture and subsistence pastoral farming. The country was divided into four regions: northern Somalia or Migiurtinia with an arid climate and little vegetation, central Somalia or Mudug, equally arid and with a nomadic population, central-southern Somalia, which had the most, developed agricultural and animal husbandry sectors and southern Somalia or lower Juba with thick vegetation and very rich fauna. One of the greatest problems was the shortage of water. Water reserves were limited to the area between the Jubba and Shabelle Rivers and outside of this area there were only wells [111]. Agriculture is by far the dominant sector in the economy: it contributes to more than 65% of GDP. The nomadic livestock sub-sector accounts for more than 44% of agricultural GDP and 50% of total agricultural employment. Nomadic livestock is the main source of Somali livelihoods. In addition to subsistence of nomad and agro-pastoralists, it also contributes about 80% of the exports and is thus the main source of the country's foreign currency earnings [112]. The Somali economy is the only one in the world where over half of the population is dependent on nomadic pastoralism [113]. The livestock sector is central to the economic and cultural life of the Somali people. Burao and Galkayo are the largest livestock markets in the Horn of Africa especially for export sheep and goats from the Somali region of Ethiopia and parts of southern Somalia [114]. The inter-riverine region which is the fertile valley that lies between the Shebelle river in the north, the Ethiopian border in the west, and the Indian ocean in the east has over fourteen ecological regions providing four modes of livelihood: agriculture, pastoralism, agro-pastoralism and trade [115].

The pastoral system is characterized by herds or flocks that are constantly moved in search of water and pasture, as the season progress. Moreover, in contrast to most pastoral systems, which are normally devoted to household subsistence, the Somali system is traditionally oriented towards trade and export [116]. After the collapse of the government in 1991, the Somali economy became entirely unofficial; however it has proven to function effectively and the livestock trade shows considerable resilience, in spite of total absence of formal institutions [117]. Livestock export in Somalia revolves around three supply chains, two of which supply the Arabian Peninsula with mostly small ruminants and cattle by sea, and the third which supplies the Kenyan market with cattle overland. [118]. [Strangio, [111] has mentioned two justifications why most of Somali population is nomadic pastoralists: first, their origins with strong cultural nomadic-pastoralist bases. Second, the climatic and environmental conditions that favour a nomadic-pastoralist economy rather than a settled agricultural economy.

5.8. Farming systems in South Sudan

Most of South Sudan country has a semi-humid climate, with annual rainfall ranging from 200-2200 mm. Rainfall is seasonal, from April to December and causes seasonal flooding of floodplains. The seasonal climate patterns cause cyclic relations in the ecosystem and hence determine land use patterns of cultivation, livestock grazing and fisheries [119].

Altitudes in Southern Sudan range from 600 to 3000 meters above sea level. Temperatures are typically above 25° C and can rise above 35° C, particularly during the dry season, which lasts during January to April [120].

South Sudan's diverse ecology provides a growing season ranging from 280-300 days in the south-western parts to 130-150 days per annum in the northern states due to bimodal and unimodal rainfall regimes. The bimodal areas cover much of Greater Equatoria, while the rest of the country has a unimodal regime [121]. There are five ecological zones in South Sudan, namely; the savannah region, the flood region, the Montane forest zone, the semi-desert zone and the lowland forest zone. The savannah region is divided into low rainfall woodland savannah zone and high rainfall woodland savannah zone. The woodland savannah region is the largest ecological region in South Sudan. The flood region includes the Suds and toic. The Suds covers about 57,000 km^2 and it is one of the largest floodplains in Africa. It is an important breeding area for Nile ecosystem fish species. Toic are areas subject to seasonal flooding by spill-water from rivers and watercourses where the soil retains sufficient moisture throughout the dry season to support grasses. It is of special importance for dry season grazing by both livestock and wildlife and is critical for the country's pastoralists. The semi-desert is the extreme southeast in and around the Ilemi Triangle; it is an extension of the northeastern Kenya semi-arid zone and shares much of the fauna and flora from that region. The lowland forest is the northernmost extension of the Congo Basin forests [122], [120]. However, the Livelihood Profile Project divided the country into seven livelihood zones that are defined based on climate conditions and farming systems: Eastern Flood Plains, Greenbelt, Hills and Mountains, Ironstone Plateau, Nile-Sobat Rivers, Pastoral and Western Flood Plains [121]. The country is naturally endowed with agricultural potential given its favourable soil, water and climatic conditions. It is estimated that about 70% of total land area is suitable for producing a wide range of agricultural products, including annual crops such as grains, vegetables, tree crops such as coffee, tea and fruits, livestock, fishery and various forest products [122].

The dominant land use in terms of land occupation is livestock keeping, which is practiced throughout almost all the country, but practically in dryer areas with better grass quality and lower livestock parasite occurrence. The vast forested areas provide, apart from food for livestock, timber, fuel wood, charcoal and non-timber forest products including food plants, medicines and bush meat. Most of the rural population practice cultivation. In the northern part of the country, sorghum, sesame and groundnut are the principal crops. Many farmers have little access to inputs and credits [119]. Crop production is mostly conducted on small, hand cultivated plots farmed by women-headed households. Sorghum is the main cultivated crop, in addition to maize, millet and rice. Other crops such as groundnut, cassava, sweat potato, vegetables and sesame are also grown. Rainfed mechanized cereal production is practiced on large scale in the Upper Nile state [123]. Crop production systems remain primarily subsistence in nature and crop yield is low. Less than 4% of the total land (about 2.7 million ha) is currently cultivated while more than 80% is still under natural vegetation [122]. According to UNEP [124] livestock rearing may be categorized into three systems: (a) nomadic, based on herding of cattle, camels, sheep, and goats (b) semi-nomadic agro-pastoralist, combining the herding of cattle and some sheep with cultivation (c) sedentary system, where cattle and small livestock are reared in close proximity to villages [124]. Livestock is mainly perceived as a store of value by many livestock keepers, and production of livestock products (butter, milk, meat and hide) is low [119]. Livestock are very important assets throughout the country, the main species being cattle, goats and sheep. The sale of livestock especially small ruminants, offer significant income generation opportunities for both transhumant pastoralists and sedentary livestock rearers [123].

5.9. Farming systems in Sudan

Geographically, four major regions are characterizing Sudan: the Northern, Western, Eastern regions and the Central Clay Plains. The northern region lies between the Egyptian borders and Khartoum. It consists of two distinct areas: the desert and the Nile Valley. The Nile River provides an alluvial strip of habitable land some 2 km wide, whose productivity depends on the annual floods. Western Sudan is a generic term describing Darfur and Kordofan, amounting to about 850,000 km^2. Its dominant feature is the absence of perennial streams, and people and animals must remain within reach of permanent wells. Consequently, the population is sparse and unevenly distributed. Eastern Sudan is located northeast of the Central Clay Plains.

It is divided between desert and semi-desert and includes Al Butanah grazing area (stretches between Khartoum and Kassala), the Qash Delta, the Red Sea Hills and the coastal plains. The Central Clay Plains stretch eastward from Nuba Mountain to the Ethiopian frontier, broken only by Ingessana Hills in Blue Nile state. These plains are productive and provide the main support of the national economy [125].

Sudan is a country of fragile eco-systems, frequent droughts, and as a result, pressing challenges to address the national priorities of food security, water supply and public health. An examination of Sudan's ecological zones indicates that the majority of its land is quite vulnerable to changes in temperature and precipitation [126]. With the secession of South Sudan, the ecology of Sudan has shifted towards a predominantly dry lands environment. The rainy season lasts less than two months in the north and extends up to four months further south. This extreme rainfall variability over time and space has a remarkable impact on vegetation, especially in more arid areas.

The country's land and water resources can be classified into four major ecological regions: (a) arid and semi-arid ecosystems, which occur in the northern and central parts of the country. Summer temperatures can often exceed 43°C, and sandstorms blow across the Sahara from April to September, with an average rainfall of 0 – 300 mm. the soil is generally poorly developed due to little rainfall and lack of vegetation. However, in some parts of this zone (Northern Darfur state), during winter times, the air may cool down at night sufficiently to form dew, allowing certain grasses, called *'gizu'* to grow after rain. Irrigated agriculture is practiced along the Nile banks, apart from that pastoralism and agro-pastoralism are also practiced. Among the challenges are desertification and severe soil degradation; (b) low rainfall savannah (sand) are typified by low rainfall (300-400 mm) and the prevalence of sandy soils. The zone is devoted to traditional rainfed farming and pastoral systems, (c) low rainfall savannah (clay) which are typified by low rainfall (400-900 mm) and clay soils, rainfed farming and pastoral livelihoods are the major farming systems in this zone. Both savannah types are characterized by a mixture of grasses and trees; (d) high rainfall savannah which is characterized by moderately high rainfall (800-1300 mm) and it represents area bordering the country of South Sudan including some parts from South Darfur and Nuba Mountains [127], [128], [125].

Based on FAO, report [125], [16], and FAO report on Nile Basin [99], the following farming systems are practiced in Sudan:

1. Irrigated farming systems: it occupies about 1, 86 million ha, dominated by gravity-supply type of irrigation. These are originally owned and managed by the public sector. The schemes are cultivated by thousands of tenant farmers and the main crops are cotton, wheat, groundnut and sorghum. Apart from these public large schemes, pump irrigation is also practiced. Performance problems in irrigation schemes include: inefficient water management; non-collection of water charges and land use fees; low productivity; large debt burdens; unsettled land rights; and inadequate financial and marketing services. On commercial basis, sugarcane for sugar industry is also widely practiced in the fertile delta lands between the Blue and White Niles. On small scale basis, there are considerable numbers of farmers practicing traditional irrigated farming along the Nile banks all over the Sudan.

2. **Traditional farming systems:** this is practiced in an area account for about 8.4 million ha where there is moderately sufficient rainfall. Over 70% of the population depends on this system for their livelihoods on crop production or livestock husbandry or both. It is dominated by small scale famers who typically live in conditions of persistence poverty and are reliant on rain-fed and traditional agricultural practices. They face challenges of climate change, droughts, rainfall variability, land degradation, desertification, low productivity and persistent food insecurity. This system includes millions of small scale subsistence farmers, who grow sorghum, millet, maize, sesame, and groundnut. They mainly depend on family labour and use of traditional tools.

3. **Rain-fed commercial semi-mechanized farming systems**: this has been developed on generally alkaline clay soils and loams. It is found in states of Gadaref, Blue Nile, White Nile, Sennar, and Southern Kordofan. These are mostly owned and managed by the private sector. Unfortunately, this system had been perceived as one of the wrong policies in the history of agriculture in Sudan. For example, Sulieman and Buchroithner [129] claimed that it has been the main contributing factor to deforestation and land degradation [129]. On the other hand, Sulieman and Elagib, [130] reported that, in recent decades, pastoralism has been in decline because of threats posed by rapid encroachment of mechanized rain-fed agriculture, human population growth and other human activities that force extensive livestock production to shift to areas of increasing marginal productivity [130]. However, the changes in land use practices have brought nomads/pastoralists into conflict with farmers both on mechanized schemes and on traditional farms [131].

4. **Pastoral farming systems**: these are entirely raised on natural rangelands and are mainly semi-nomadic; however, nomadic and transhumance are also there to strategically utilize the available rangeland resources. Households move with their animals and spend the rainy season in the northern semiarid zone at places of availability of both pasture and water and where they can avoid biting insects and the muddy conditions. In the dry season, they move to the south, the savannah areas. In the Eastern and the Central areas of Sudan the migration is towards the Nile during the rainy season and back during the dry season. Movement is usually practiced along the livestock routes or corridors, traditionally known as 'Masarat' or Maraheel'. The major challenges include: shortage of water, animal feed, diseases, and horizontal expansion of mechanized, rain-fed cultivation, land degradation, conflicts, and lack of capital and poor marketing services of the livestock products. It has been emphasized by Behnke [132] that unlike other countries in the region, pastoralism is not merely an occupation of the peripheral areas of Sudan; pastoral rangelands are distributed throughout Sudan, even in Khartoum State itself, and are the backbone of livestock production in Sudan [132].

5. **Agro-pastoral farming systems:** it is practiced under conditions fairly similar to those of pastoral systems, but differs in the slightly available water sources and growing of some crops. Crops are grown, but play a less significant role than in most other farming systems as moisture stress is a limiting critical factor on crop production. Livestock production is the main source of income and food. Among the constraints are some socio-cultural practices, land degradation, land tenure problems, diseases, lack of security and low rainfall.

5.10. Farming systems in Uganda

Uganda lies between latitudes 4° N to 1° S and longitude 29° E and 36° E. Although temperature variations may be significant, especially over high ground areas in western, eastern, south western and parts of northern Uganda, rainfall, like in many tropical areas largely determines the climatic sub-regions (agro-climatic zones) of the country. It also determines the spatial patterns of natural resources and land use activities [133] A large part of Uganda consists of a plateau, lying between 1000 and 2500 above sea level. Temperatures are moderate, between 15° and 30°C. Precipitation varies from 750 mm to 1500 mm. Due to climate change; the onset of the rainy season is increasingly unreliable, and rainfall distribution is more uneven with erratic, heavy rainfall events [134]. It is endowed with large fresh water resources and a high agricultural potential. The population, young and predominantly rural, is mostly engaged in subsistence rainfed farming. [135]. The main water bodies are Lake Victoria, Lake Albert, Lake Kyoga, Lakes Edward and George. The main rivers include the Victoria Nile, and the Albert Nile [136]. The Nile Victoria divides the country in two parts, flowing from Lake Victoria at Jinja through Lake Kyoga to the northern tip of Lake Albert, and then north to South Sudan. The climate is tropical but mild because of the generally high altitude. The temperature ranges from about 16° to 29° C, with 1000 mm or more rainfall over most of the country. In the extreme north-east, in Karamoja, there is small zone with less than 500 mm of rainfall. Rainfall is bi-modal in the central and western regions, and mono-modal in the northern and eastern regions. In the central and western regions, the months of December to mid-February and June to mid-August are usually dry periods [135]. Soil fertility varies according to the level of rainfall. The land is generally fertile in the central and western regions and becomes less fertile as one move to the east and the north [137].

CIAT [138] had defined, delineated and characterized about 33 agro-ecological zones for Uganda, and then they have been aggregated into 14 zones. A number of classifications of agricultural production systems have been developed for Uganda [138]. For example, five systems have been distinguished by NEMA [134]. These systems are: northern and eastern cereal-cotton-cattle, intensive banana-coffee, western banana-coffee-cattle, west Nile cereal-cassava-tobacco, and Kigezi afromontane.

Based on these zones, Mwebaze [139]; Kabeere and Wulff [140]; Ronner and Giller [141]; classified the farming systems into seven as follows:

1. **The banana-coffee system:** in this system, rainfall is evenly distributed (1000-1500) on soils of medium to high productivity. The areas cultivated per capita are small, under one hectare. The main crops are banana, coffee, maize, and sweet potatoes.

2. **The banana-millet-cotton system**: rainfall for this system is less stable than for the banana-coffee system, so there is greater reliance on annual food crops (millet, sorghum and maize). In the drier areas, livestock is a main activity.

3. **The montane system:** it is found at higher elevations between 1500-1750 metres above sea level. The area receives high and effective rainfall and cloud cover. Crops grown are banana, sweet potatoes, cassava and Irish potatoes. Arabica coffee is prevalent at above 1600 metres. Some temperate crops like wheat and barley are grown.

4. **The teso system:** the area receives bimodal rainfall on sandy loams of medium to low fertility. The dry season is longer from December to March. The vegetation is moist and grass savannas, short grassland which is ideal for grazing. Main crops are millet, maize, sorghum, cotton and oil seed crops. Mixed agriculture (crops and livestock) is practiced. Livestock are kept extensively in those areas which are tsetse-fly free.

5. **The northern system**: the rainfall in this system is less pronouncedly bimodal with about 800 mm annually. The dry season is so severe that drought tolerant annuals are cultivated; these include finger millet, sesame, cassava, and sorghum. Tobacco and cotton are the major cash crops. The grassland is short and communal grazing abounds. This area is well known for its pastoral system with semi-nomadic cattle herding.

6. **The West Nile system:** the rainfall pattern resembles that of the northern system, with more rain at higher altitudes. Mixed cropping is common with a wide variety of crops. Livestock activities are limited by the presence of tsetse fly. Tobacco and cotton are also the major cash crops.

7. **The pastoral system:** this system covers some districts in the north-east, parts of Western and Central districts. Annual rainfall is low (less than 100 mm). The system is characterized by short grassland where pastoralism prevails with nomadic extensive grazing. The livestock production systems are generally two main groups according to Mbuza [142], they are: the traditional systems and the improved systems. However, according to Mwebaze [139] and based on the grazing methods, there are about seven livestock production systems, namely:

    a. **Communal pastoral systems:** this is prevalent in the south west of the country, in the central areas, and in the north and north east. Indigenous breeds of cattle, goats and sheep are raised depending on natural grazing. Among the factors limiting production: water scarcity, sparse population, low vegetation with low grass quality, low literacy rate, lack of effective extension systems, breeds are genetically poor and wide spread diseases.

    b. **Agro-pastoral system**: are sedentary farmers who grow food crops both for subsistence and sale, while keeping some livestock which graze on communal land, fallows and on crop residues. Nowadays, with the increase in population and land pressure, this system evolves into mixed farming. Among the limiting factors: little control over land, crop residues and other feed sources, high mortality rates, diseases, uncontrolled mating, and reduced grazing time.

    c. **Tethering system:** this semi-intensive system, where livestock are restrained by a rope, is common in urban, peri-urban and intensively cultivated areas where herd size is small. Crop production is the farmers' main activity. Among the limiting factors are: losses due to diseases and predators, scarcity of water and veterinary and extension services, negative genetic selection, no fodder banks and environmental degradation.

    d. **Fenced dairy farming system:** this is an intensive or semi-intensive system of dairy farmers where farmers use part or all of their land to plant or improve pastures and grow fodder. They may also buy concentrates. This system is found in south Western Uganda Central and south eastern parts of Uganda.

e. **Zero grazing system**: this is on an increase in and around urban areas where land is scarce but there is good market for milk and other livestock products. It is not traditional and is intensive, livestock is continuously housed and owners have to establish fodder gardens to provide green forage. The main sources of feed are fodder, crop residues, domestic wastes, and agro-industrial by-products. Among the limitations are: labour intensive, forage crops occupy the land at the expense of food crops, high capital outlay, high cost of feeds, weed problems and difficulty of providing water.

f. **Mixed farming system:** in Uganda, it is common to combine livestock and crop production, the two enterprises are complementary. Crops are the main agricultural activity. Livestock are kept for draught, milk and/or meat for sale. It is common smallholder dairy system in the south-east, central and south eastern parts.

6. Conclusions, recommendations and policy implications

6.1. Conclusions

It is apparent from the different parts covered by this chapter that any single agricultural system is open to both nature and the society existing around it or more specifically to both the bio-physical and socio-economic entities.

The analysis of farming systems is perhaps one of the instruments adopted to study, among others, agricultural policies at national, regional and even at global level. It will also provide a framework of analysis to consider the different agro-ecological zones available in one setting i.e. a country or a region as well the socio-economic characterization of the population and their livelihoods. Within that framework changes in policies and other factors that negatively or positively impact levels of agricultural production in each system, could be foreseen. However, both homogeneity and heterogeneity of bio-physical and socio-economic constituents of each agricultural system could be identified and used as variables within that analysis. In this regard, it may worth mentioning that, each farming system has its own characteristics in terms of its physical environments, market linkages, household traits and other social and economic characteristics. These characteristics of the analyzed farming system will act as pillars upon which any changes in production within the boundaries of that system are possible.

From socio-economic point of view, in addition to the available set of secondary data, about each farming system, detailed set of primary data, as well, will be needed to help knowing the developments and opportunities within the farming system under investigation. However, the detailed data will make possible grouping of the producers in each system into socio-economic strata within their corresponding livelihood systems. In addition to that, the socio-economic characterization of each system will be identified and knowing this at national level allows reviewing of the possible policy change or impact. At the regional level, the aggregated policy change will be examined for further adjustments and harmonization. Knowing of these detailed characteristics of each farming system is required urgently under the current global, regional and national concerns, such as climatic risks, poverty levels, environmental risks. As well, identifying the limiting constraints for the performance of each farming system is also valuable particularly for tailoring appropriate interventions that can bring some change in terms of practical solutions. Of course, tailoring of these interventions as practical solutions would mean considering the basic idea of being profitable, co-efficient and sustainable innovative solutions.

The Horn of Africa region is well endowed with natural resources, yet its countries still facing severe incidences of famine and poverty, compared with other developing regions. Agricultural sector in these countries represent the backbone of their economies, yet still unable to perform efficiently. The analysis of the farming systems existing in the region and its countries indicates the following remarks:

1. The Horn of Africa is diverse; its countries share specific characteristics. Their populations are divided along ethnic, border and religious lines. The region is the least developed food unsecured, poorest and conflicted region in the world.

2. Some of the studied farming systems lie within broad altitudinal and latitudinal ranges that lead to a wide agro-ecological categorization.

3. Most of them are heterogynous in nature, some are sparsely populated, others are densely, and ranging from highlands to low and dry land environments, ranging from purely irrigated to mostly rain-fed systems and some are specialized in terms of the produce, while others are run on multi-products nature.

4. They are characterized by duality in terms of presence of traditional practice alongside somewhat modernized practices. Some of them are purely subsistence while others are market oriented. There is weak engagement of the private sector in this sector in general.

5. Some of them although found in higher potential areas but they show very poor agricultural performance compared to those at lower potential areas. This implies that the agro-ecological zone though very important, but not the only factor.

6. Some of them are found at poor agro-ecological conditions while others are at good and relatively good agro-ecological conditions.

7. There are no clear linkages between these farming systems and the input and output markets. There are also weak land management practices together with weak extension service delivery.

8. For both crop and animal agriculture, there seem clear symptoms of poverty incidences particularly for smallholder producers.

9. Farming systems existing in the region are highly dynamic which necessitates the need for understanding them carefully in order to adopt sustainable interventions that may bring positive change for the people and their livelihoods.

10. Productivity levels for most of the studied farming systems are very low and they lack understanding about how to cater for sustainability and efficient use of resources.

http://dx.doi.org/10.5772/60012

11. Regarding the animal-based farming systems, among their challenges are: rangeland degradation, water and feed scarcity, poor infrastructures, animal diseases, agricultural expansion, conflicts, land tenure problems, genetically poor breeds, low rainfall and lack of effective extension services.

12. Regarding the crop-based farming systems, among the challenges they face are: soil-related problems, land tenure problems, shortage of inputs, chronic poverty, managerial and financial problems, problems related to marketing and lack of infra-structure, high costs of production, and low production levels.

6.2. Recommendations and policy implications

Based on the preceding analysis of farming system in the region, the following recommendations and priority interventions could be indicated to help overcoming some of the challenges faced these systems:

1. Taking into account the set of crises that put all the countries of the Horn (IGAD countries), its natures and how it affect these countries, there is an urgent need for regional approach to tackle these problems, or more specifically regional integration instruments to enhance the development of the poorest economies, to build mechanisms for conflict management and to build resilience of the countries, as well as that of their communities particularly those at marginalized borderlands. These development approaches are in fact possible given the fact that now the Intergovernmental Authority on Development is doing more that that including coordination and implementation of projects that will surely foster development at national and regional levels.

2. There is a need for promotion of social and economic dimensions of the agricultural systems, while dealing with the productive and bio-physical environment. This may be crucial given that the historical definitions of agro-ecology are embedded within agronomy and ecology without inclusion of the socio-economic dimensions. There is also need for incorporation of the principles of agro-ecological approaches into our education curricula and our development interventions.

3. As poverty alleviating strategy, one could recommend diversification out of one produce into mixture of high value crops with livestock particularly in the so called high potential agro-climatic zones. For those found at low potential zones, resorting to off-farm employment can be one of the solutions out of poverty trap.

4. With more poverty incidences prevailing at the small scale traditional farming systems, there is urgent need for adoption and implementation of an appropriate set of policies at national and regional levels, together with implementation of ecologically sound interventions as scaling-up of agro-ecosystem thinking towards more sustainable agricultural systems.

5. Problems related to soils could be dealt with through encouragement of more sustainable forms of land management practices such as use of natural fertilization, inter-cropping and conserving agriculture.

6. It is crucial to promote the use of appropriate extension approaches including the encouragement of the private sector to invest in this sector.

7. Regarding the pastoral systems, there is need for proper investments with institutional support from the public sector in terms of veterinary services, value addition, and marketing and information facilities. In the region, many studies had shown the significant importance of these systems in the national economies of most of the countries in the region, but more efforts in terms of awareness as these facts are not fully appreciated by governments, policy makers and development partners.

8. For small scale producers whether they are farmers or pastoralists, there is need to advocate for development interventions that are imposed which will make sustainability impossible, so there is need for interventions that are based on participatory development approach. There is also need for adoption of sustainable natural resource management as viable policy option.

9. To tackle problems of marketing and inputs especially for small scale producers (farmers and pastoralists) critically, one of the solutions may be the formation of organizations, policy forums for the producers to help in catering for provision of those services and to advocate for their rights.

10. Support enhancement of policies, regulations and frameworks that foster accessibility to domestic, regional and international markets

11. To enhance solutions of food insecurity and poverty prevalence, problems facing the large scale irrigated farming systems need to be tackled, below are some examples:

    a. The management-related problems could be dealt with through inclusion of farmers in the management board of these schemes.

    b. The water-related ones can be solved through formation of water users associations.

    c. The problems related to land issues could be seen under formulation of proper land policies.

    d. Soil-related issues such as decline in fertility is better be managed through natural fertilization and inter-cropping.

12. Given the expected incidences resulting from the current risks (environmental, climatic, market,), depending on indigenous resilience of the farming systems alone is not enough to create a resilient and transformable agricultural systems. For that, Key technical solutions in terms of sustainable intensifications, policy and market development are helpful to overcome these difficulties.

13. To promote structural transformative change within the studied farming systems, formulation of appropriate set of policies of land, water and rangelands for sustainable co-efficient use of resources.

http://dx.doi.org/10.5772/60012

Author details

Osman Babikir[1*], Solomon Muchina[2], Ameha Sebsibe[3], Adan Bika[4], Agol Kwai[5], Caroline Agosa[6], George Obhai[7] and Samuel Wakhusama[8]

*Address all correspondence to: osman.babikir@igad.int

1 Social Economics, Policy and Marketing, IGAD Centre for Pastoral Areas and Livestock Development (ICPALD), Nairobi, Kenya

2 IGAD Centre for Pastoral Areas and Livestock Development (ICPALD), Nairobi, Kenya

3 Department for Livestock and Fisheries, ICPALD, Nairobi, Kenya

4 Department for Dryland, Development and Climate Change adaptation, Nairobi, Kenya

5 Department for Animal Health, ICPALD, Nairobi, Kenya

6 Department for Agro-climate, ICPALD, Nairobi, Kenya

7 Department for Monitoring and Evaluation, ICPALD, Nairobi, Kenya

8 Department for Veterinary Governance Project for IGAD Region, Nairobi, Kenya

References

[1] Gomeiro et *al.* Facing complexities on agro-ecosystems: a new approach to farming system analysis. International Journal of Agricultural Resources, Governance and Ecology (2006). 5(2/3), 116-144.

[2] IIASA and FAO. Global Agro-ecological Zones (GAEZ v3.0). IIASA, Laxenburg, Austria and FAO, Rome, Italy (2012).

[3] Gliessman, S. R. Agro-ecosystem sustainability: developing practical strategies. (2000) CRC Press: Boca Raton FL, USA.

[4] Gliessman, S. R Agroecology and sustainability, INTECOL Symposium 8.2. Florence, Italy, (1998). http://www.agroecology.org/documents/Intecolrep.pdf

[5] Desai, B. and Pujari, B. T. Sustainable agriculture: a vision for future New India Publishing (2007)

[6] Mateo N and Rodomiro, O. Resource use efficiency revisited. In: Clare H and Paul N Eco-efficiency: from vision to reality Colombia, CIAT Publications (2013), 1-17.

[7] Koohafkan, P., Altieri, M. and Gimenez, E. Green agriculture: foundations for biodiversity, resilient and productive agricultural systems. International Journal of Agricultural Sustainability, (2011).

[8] Francis C. et *al.* Agro-ecology: the ecology of food systems. Journal of Sustainable Agriculture (2003). 22(3), 99-113.

[9] www.agroecology.org, accessed 2015

[10] Gliessman, S. R. Economic and ecological factors in designing and managing sustainable agro-ecosystems. In: Edens, T and Fridgen C. (ed.). Sustainable agriculture and integrated farming systems Michigan State University Press, East Lansing, Michigan. (1985) 56-63.

[11] Andriesse, W et *al.* The role of agriculture in achieving MDG1- A review of the leading reports. Wageningen International, Wageningen (2007).

[12] Garrity, D et *al.* Understanding African farming systems science and policy implications. Paper prepared for Food security in Africa: bridging research and practice.

[13] Giessen E Horn of Africa Environmental security assessment. Institute of Environmental Security Assessment Report, The Hague, the Netherlands (2011).

[14] Schiere, J. et al Evolution of farming systems and system philosophy. Systems research and behavioral sciences, 1999, 16, 375-390.

[15] Dillon J. and John, D. Farm management for Asia: a system approach, Publication of Food and Agriculture Organization of the United Nations, FAO, Rome (1997).

[16] Dixon, J et al. Farming systems and poverty improving farmers' livelihoods in a changing world, FAO and World Bank, Rome and Washington, (2001).

[17] Okigbo, B. Evaluation for plant interactions and productivity in complex mixtures as a basis for improved cropping systems design. Proceeding International workshop on intercropping, Hyderbad, India, (1979), 350-356.

[18] Rouw, D. Rice, weeds and shifting in a tropical rainforest, a study of vegetation dynamic. Doctorate thesis, Agricultural University of Wageningen, (1991).

[19] Rosen, R. Life itself: a comprehensive inquiry into the nature, origin, and fabrication of life. New York, Columbia University Press, (1991).

[20] Altieri, M. Agroecology: the science of natural resource management for poor farmers in marginal environments. Agriculture, ecosystems and environments, (2002). 93, 1-24.

[21] Ker, A. farming systems of the African savanna a continent in crisis. International Development Research Centre, (1995).

[22] Norman, D. and Malton, P. Agricultural systems research and technical change. In: J. P. Colin, E. W. Crawford (eds) Research on agricultural systems. Accomplishments, perspectives and issues. Nova Science Publishers, Inc, (2000), 17-47.

[23] Spedding, C. The biology of agricultural systems. In: Biological Systems. Academic Press, London. (1975), 42-65.

[24] Ruthenberg, H. Farming systems in the tropic. Third edition, Clarendon Press, Oxford, UK (1980)

[25] Fresco, L and Westphal, E. A hierarchical classification of farm systems. Exp. Agriculture, (1988), 24, 399-419.

[26] Duckham, A and Masefield, G. Farming systems of the world. Landon, Chatto and Windus. (1970).

[27] Grigg, D. Agricultural systems of systems of the world an evolutionary approach. Cambridge, Cambridge University Press. (1974).

[28] Robinson, T.P., Thornton P.K., Franceschini, G., Kruska, R.L., Chiozza, F., Notenbaert, A., Cecchi, G., Herrero, M., Epprecht, M., Fritz, S., You, L., Conchedda, G. & See, L. Global livestock production systems. Rome, Food and Agriculture Organization of the United Nations (FAO) and International Livestock Research Institute (ILRI).

[29] Doppler, W. Farming and rural systems – state of the art in research and development. In: Doppler, W. and J. Calatrava, (eds.): Technical and social systems approaches for sustainable rural development. Granada, (2000). 1-20. ifsa.boku.ac.at

[30] Power, A. Ecosystem services and agriculture: tradeoffs and synergies. Philosophy Transaction Royal Society, (2010). 365, 2959-2971.

[31] Guyer, J and P. Peters. Introduction: conceptualizing the household. Development and Change. (1978), 18, 197-214.

[32] Mengisteab, K. The Horn of Africa. Polity Press, Cambridge, UK, (2014).

[33] Tekle, A. International relations in the Horn of Africa (1991-1996). Review of African Political Economy, (1996), 70, 499-509.

[34] Degu, W. The state, the crisis of state institutions and refugee migration in the Horn of Africa: the cases of Ethiopia, Sudan and Somalia. Dissertation, Faculty of Social and Behavioural Sciences, University of Amsterdam, Netherlands, (2002).

[35] Lewis, M. Languages of the world (16th edn.), Dallas TX: SIL International; (2009), http://www.ethnologue.com/country_index.asp?place=Africa

[36] Michalopoulas, S and Elias P. The long run effects of the scramble for Africa, National Bureau of Economics Research, Working Paper 17620, (2011).

[37] Samatar, M and Mackaka, W. Conflict and peace in the Horn of Africa: A regional approach. In: Quest for a culture for peace in the IGAD region, Nairobi, Heinrich Boll Foundation, (2006), 26-55.

[38] Feyissa, D and Hoehne M. State borders and borderlands as resources. In: Feyissa, D and Hoehne, M (eds). Borders and borderlands as resources in the Horn of Africa. James Currey, UK, (2010).

[39] Asiwaju, A. and Nugent, P. Introduction: The paradox of African boundaries. In: A. L. Asiwaju and P. Nugent (eds.), African boundaries: Barriers, conduits and opportunities, London Printer, (1996), 1-17.

[40] African Development Bank (AfDB), the African Union Commission (AUC) and the United Nations Economic Commission for Africa (UNECA) African Statistical Yearbook, (2014).

[41] Abdi, A and Aragie, E. Economic growth in the Horn of Africa: Identifying principal drivers and determinants. The Horn Economic and Social Policy Institute (HESPI), Policy Paper No. 12/3, (2012).

[42] Healy, S. Hostage to conflict prospects for building regional economic cooperation in the Horn of Africa. Chatham House Report, London, (2011).

[43] United Nations Development Program (UNDP). Human development Report, 2014.

[44] Dilly, M. *et al.* Natural disaster hotspots: A global risk analysis. The World Bank and Collumbia University, (2005).

[45] Oxfam Briefing on the Horn of Africa drought: climate change and future impacts on food security. Briefing report, (2011).

[46] USAID Djibouti biodiversity and tropical forests: assessment. (1998).

[47] www.immigration-usa.com/world_fact_book_2012/djibouti/index.html

[48] The World Travel Guide (2012). Djibouti weather, climate and geography. www.worldtravelguide.net/

[49] International Monetary Fund (IMF) Report. Djibouti: poverty reduction strategy paper. International Monetary Fund, Washington, DC, (2009).

[50] Brass, J. Djibouti's unusual resource curse. Journal of Modern African Studies, Cambridge University Press, (2008), 46(4), 523-545.

[51] International Fund for Agricultural Development (IFAD). Eritrea: country strategic opportunities paper, (2006).

[52] en.wikipedia.org/wiki/Eritrea

[53] African Development Bank, African Economic Outlook Report, (2014). www.africaneconomicoutlook.org

[54] African Development Bank, Interim country strategy paper for Eritrea, (2009).

[55] Food and Agriculture Organization (FAO) Country pasture/forage resources profiles, Ethiopia, Mengistu, A. (2006).

[56] International Monetary Fund Report (IMF). The Federal Democratic Republic of Ethiopia: poverty reduction strategy paper: growth and transformation plan 2010/2011-2014/2015 – volume one (2011).

[57] en.wikipedia.org/wiki/Kenya

[58] Library of Congress, Federal Research division, Country Profile, Kenya (2007).

[59] International Fund for Agricultural Development (IFAD). Republic of Kenya Country Strategic Opportunities Programme, Rome, Italy (2013).

[60] http://www.americaregistry.com/, accessed February 2015

[61] Library of Congress, Federal Research Division, Country Profile, Somalia, (1992)

[62] United Nations Development Programme (UNDP). Somalia Human Development Report Empowering youth for development report, (2012).

[63] African Development Bank (AfDB) Somalia Addendum to Country Brief 2013-2015, (2013).

[64] en.wikipedia.org/wiki/South Sudan

[65] Lovell-Horaire, S and Lovell-Horaire, M, South Sudan. Bradt Travel Guides Ltd, (2013).

[66] Youngo, B. and Michigan, F. Peace dividend and the millennium development goals in Southern Sudan,

[67] en.wikipedia.org/wiki/Sudan

[68] Willis, J. and Justin, W. Land and Water. In: Ryle, J. and Justis, W, Suleiman, B, Jok, M.The Sudan Handbook, James Currey an imprint of Boydell and Brewer Ltd., UK and US. (2011).

[69] World Bank (WB). Interim Strategy Note (2014-2015) for the Republic of the Sudan. (2013).

[70] World Bank (WB). Sudan the Road to Sustainable and broad-based growth. (2009).

[71] en.wikipedia.org/wiki/Uganda

[72] IFAD (International Fund for Agricultural Development). Country Strategic Opportunities paper, (2013).

[73] Niemi, E and Manyindo, J. Economic importance of goods and services derived from dryland ecosystems in the IGAD region: Case studies. IUCN, Nairobi, (2010).

[74] Mortimore, M. Dryland opportunities: A new paradigm for people, ecosystems and development. IUCN, (2009).

[75] Flintan, F, Behnke, R and Neely, C. Natural resource management in the drylands in the Horn of Africa. Brief prepared by a Technical Consortium hosted by CGIAR in partnership with FAO Investment Centre. Technical Brief 1. Nairobi, International Livestock Research Institute. (2103).

[76] Knips, V. Review of livestock sector in the Horn of Africa. Former IGAD-LPI project (FAO), (2004).

[77] Shazali, S and Abdel Ghafar, M. Pastoral land tenure and agricultural expansion: Sudan and the Horn of Africa. IIED Issue, paper No. 85, (1999).

[78] Ahmed, A., Livelihood and resource competition, Sudan. In: Mohammed, S, Ton, D. and Abdel Ghafar, M (eds.). African pastoarlism conflict, institutions and government, Pluto Press, London, OSSREA Publications, (2001).

[79] Standford, S Management of pastoral development in the third world. London, ODI, (1983).

[80] African Development Bank (AfDB), Regional study on sustainable livestock development in the Horn of Africa. African Development Group, LOG Associates, (2010).

[81] Otte, M and P. Chilonda. Cattle and small ruminant production systems in Sub-Saharan Africa – A systematic review, Food and Agriculture Organization (FAO), Rome (2002).

[82] Ginkel, M, et al. An integrated agro-ecosystem and livelihood systems approach for the poor and vulnerable in dry areas. Springerlink.com (2013).

[83] Fowler, C. and Hodgkin, T. Plant genetic resources for food and agriculture: Assessing global availability. Annual Review Environmental Resources, (2004), 29, 143-179.

[84] Masetre, F, et al. Plant species richness and ecosystem multifunctionality in global drylands. Science, (2012), 335, 214-218.

[85] Harvey, C. et al. Conservation value of dispersed tree cover threatened by pasture management. Forest Ecology and Management, (2011), 261 (10), 1664-1674.

[86] Roetter, R. et al. Systems research for optimizing future land use in South and Southeast Asia. Proceeding of the SysNet'99 International Symposium. SysNet Research Paper Series 2. Los Banos: International Rice Research Institute, (2000).

[87] Convention on Biological Diversity (CBD). Country Profile, Djibouti, (2012).

[88] United Nations Development Program (UNDP). Programme Proposal. Adaptation Fund Project, Djibouti, (2012).

[89] Food and Agriculture Organization (FAO). The FAO Component of the Consolidated Appeals, Djibouti. (2011).

[90] Brass, J. The political economy of livestock development: The case of Djibouti, IGAD-LPI working paper 02-08 (Inter-governmental Authority on Development Livestock Policy Initiative), (2007).

[91] Library of Congress, Federal Research division, Country Profile, Eritrea (2005).

[92] Frey, L, Sotoudeh, S and Stllhardt, B. A baseline survey for sustainable development of the Deki Lefay Community, Eritrea. Centre for Development and Environment, University of Berne, Swizerlands. (1997).

[93] Ghebru, B et al. Eritrea. In: Waithaca, M, Gerald C, Timothy, S, and Miriam, K, (eds.) East African Agriculture and Climate Change: A Comprehensive Analysis. The International Food Policy Research Institute Publications (IFPRI)

[94] African Development Bank (AfDB). Interim Country Strategy paper for Eritrea, (2009-2011). (2009).

[95] USAID Ethiopia biodiversity and tropical forests: 118/119 Assessment. (2008).

[96] IFAD (International Fund for Agricultural Development). The Republic of Ethiopia Country Strategic Opportunities paper, (2008).

[97] Kibret, H. Land reform revisiting the public versus private ownership controversy. Ethiopian Journal of Economics, (1998), 7(2), 45-64.

[98] Desalegn, R. Agriculture policy review. In: tesfaye, T. (eds) Digest of Ethiopia's National Policies, strategies and Programs, FSS, Addis Ababa, (2008), 129-151.

[99] Food and Agriculture Organization of the United Nations (FAO). Farming systems report: Synthesis of the country reports at the level of the Nile Basin. (2011).

[100] Makokha, M et al. Kenya. In: Waithaca, M, Gerald C, Timothy, S, and Miriam, K, (eds.) East African Agriculture and Climate Change: A Comprehensive Analysis. The International Food Policy Research Institute Publications (IFPRI)

[101] Jaetzold R. H. farm management handbook of Kenya (Vol.11), Ministry of Agriculture, Kenya and the German Agricultural Team (GAT) of the German Agency for Tropical Cooperation (GTZ), Nairobi, Kenya. (1983).

[102] USAID Kenya tropical forest and biodiversity assessment prosperity, livelihoods and conserving ecosystems (2008).

[103] Sombroek, W., Braun, H., and Van der, P. Explanatory soil map and agro-climatic zone map of Kenya. Report for National Agricultural Laboratories, Soil Survey Unit, Nairobi, Kenya, (1982).

[104] World Resource Institute. Nature's benefits in Kenya. An Atlas of ecosystems and human well-being, Washington DC and Nairobi. (2007).

[105] Barrett, C, et al. Livestock pricing in the Northern Kenyan rangelands. Journal of African Economies, (2003), 12(2), 127-155.

[106] NEMA and UNDP Fourth National Report to the Conference of Parties to the Convention on Biological Diversity. National Environmental Management Authority, Nairobi, Kenya, (2009).

[107] Neubert, S. Poverty Oriented Irrigation Policy in Kenya Empirical results and Suggestions for Reform. German Development Institute, Discussion Papers, (2007).

[108] Ngigi, S. Review of irrigation development in Kenya. In: Blank, H, Mutero, C and Murray, H (eds.). The changing face of irrigation in Kenya opportunities for anticipating changein eastern and Southern Africa, (2002).

[109] Cecchi, G et al. Geographic distribution and environmental characterization of livestock production systems in Eastern Africa. Agriculture, Ecosystems and Environment, (2010), 135, 98-110.

[110] Mati, B. et al. Assessing water availability under pastoral livestock systems in drought prone Isiolo District, Kenya. Working Paper 106, Colombo, Sri Lanka: International Water Management Institute (IWMI), (2005).

[111] Strangio, D., The reasons for underdevelopment: The case of decolonization in Somaliland, contribution to economics, Physical Verlang, Springer Company, Berlin Heidelberg, (2012).

[112] Ruggieri, C. State reconstruction and economic recovery in Somalia: An alternative option between central state and clan-based systems of governance. Journal of Middle Eastern Geopolitics, (2004), 25-38.

[113] Drydale, J. Stoics without pillows, a way forwards for the Somaliland. Haan Associates Publishing, London (2000).

[114] Food and Agriculture Organization of the United Nations (FAO) GIEWS Country Brief, Somalia, (2014).

[115] Mukhtar, H. The plight of the agro-pastoral society of Somalia. The Review of African Political Economy, (1996), 23(70), 543.

[116] Abdullahi, A. Pastoral production system in Africa: A study of nomadic household economy and livestock marketing in central Somalia. Farming systems and resource economics in the tropics, Kiel, Germany, (1990).

[117] Little, P. Somalia: Economy without state. Indiana University Press, Bloomington, Indiana, (2003).

[118] Tempria, S. et al. Mapping cattle trade routes in southern Somalia: A method for mobile livestock keeping systems. Rev. Sci. tech. off int. Epiz. (2010), 29(3), 485-495.

[119] United Nations development Programme (UNDP). Environmental impacts, risks and opportunities assessment: Natural resource management and climate change in South Sudan. Ministry of Environment, South Sudan and UNDP Report, (2012).

[120] USAID Southern Sudan environmental threats and opportunities assessment Biodiversity and tropical forest assessment. (2007).

[121] South Sudan Centre for Statistics (SSCSE). Southern Sudan livelihood profiles: A guide for humanitarian and development planning. SSCE, Save the Children UK and FEWSNET.

[122] Diao, Z. et al. Assessing agricultural potential in South Sudan. In: Monwar, B (eds), Application of geographic information systems, Intech Publishing, (2012), www.intechopen.com

[123] Food and Agriculture Organization (FAO and World Food Program (WFP). Crop and food security assessment mission to South Sudan report, (2014).

[124] United Nations Environment Programme (UNEP). Sudan post-conflict environmental assessment. Nairobi, Kenya, (2007).

[125] Food and Agriculture Organization of the United Nations (FAO). Fertilizer use by crop in Sudan, published report, Land and Water Development Division, FAO, Rome, Italy (2006).

[126] Republic of Sudan, Ministry of Environment and Physical Development (MEPD), Higher Council for Environment and Natural Resources. Sudan's first national communications under the United Nations Framework Convention on Climate Change, (2003).

[127] Noorddwijk, M. Ecology textbook for the Sudan, Khartoum University Press, (1984).

[128]] Taha, A et al. Sudan. In: Waithaca, M, Gerald C, Timothy, S, and Miriam, K, (eds.) East African Agriculture and Climate Change: A Comprehensive Analysis. The International Food Policy Research Institute Publications (IFPRI)

[129] Sulieman, H and Buchroithner, M. Degradation and abandonment of mechanized rainfed agricultural land in the southern Gadarif region, Sudan: The local farmers' perception. Land Degradation and Development, (2009), 20(2), 199-209.

[130] Sulieman, H. and Elagib, N. Implications of climate, land use and land cover changes for pastoralism in Eastern Sudan. Journal of Arid Environments, (2012), 85, 132-141.

[131] Osman, A. M. Agricultural change, land and violence: An examination of the region of Darfur, Sudan, PhD Dissertation, Food Policy and Applied Nutrition, Gerald J. and Dorothy R. Friedman School of Nutrition Science and Policy, Tufts University, Boston MA.

[132] Behnke, R and Osman, H. The contribution of livestock to the Sudanese economy, IGAD LPI working paper No. 01-12, IGAD Livestock Policy Initiative.

[133] Ejiet, J. assessment of the impact of climate change and climate variability on crop production in Uganda. Report submitted to Global Change System for Analysis, Research and Training (START) Us National Science Foundation (NFS), (2009).

[134] NEMA. State of the Environment Report for Uganda (2010). Kampala: National Environment Authority (NEMA).

[135] Food and Agriculture Organization of the United Nations (FAO). The Republic of Uganda, Nutrition Country profile, (2010).

[136] Ibale, R. Towards an appropriate management regime for the fisheries resources of Uganda. Final project report (1998), Fisheries Department, Ministry of Agriculture, Animal Industry and Fisheries, Uganda, www.unuftp.is/proj98/robbinprf.pdf

[137] UBOS and Macro International Inc. Uganda Demographic and Health Survey. Uganda Bureau, Kampala, and Macro International Inc. Calverton, Maryland, USA. (2007).

[138] International Centre for tropical Agriculture (CIAT). Uganda's Agro-ecological Zones: A guide for Planners and Policy Makers, CIAT, Kampala, Uganda, (1999).

[139] Mwebaze, S. Country pasture/forage resource profile, Uganda, (2006), www.fao.org/uganda

[140] Kabeere, F. and Wulff, E. Seed sector country profile: Uganda. Volume 1: Overview of seed supply systems and seed health issues. Copenhagen: Danish Seed Health Centre for Developing Countries, University of Copenhagen (2008).

[141] Ronner, E and Giller, K. Background information on agronomy, farming systems and ongoing projects on grain legumes in Uganda, www.n2africa.org

[142] Mbuza, F. A systems analysis of milk production system in Uganda and prospects for technology change. PhD Thesis, Australia, University of Melbourne.

Chapter 2

Characterization of Industrial Highly Organic Wastewater to Evaluate Its Potential Use as Fertilizer in Irrigation of Agricultural Land

Esteban Jácome Sandoval and
Martin H. Bremer- Bremer

Additional information is available at the end of the chapter

http://dx.doi.org/10.5772/59999

1. Introduction

According to the United Nations [1], the worldwide consumption of water for industrial use represented 20% of the total available, equivalent to 784 km^3year. The prediction is that by the year 2025, the consumption of water by industry will have increased by 50%. On the other hand, it is estimated that the wastewater produced by this sector generates around 500 million tonnes of organic matter per year [2], which should be treated adequately prior to its being released to water bodies.

On the other hand, agriculture is the productive sector with the greatest water consumption around the world, reaching 70% of the total annual withdrawal [3]. In several developing countries, approximately 95% of the extracted water is used for agricultural activities and therefore plays a key role in food production and food security [4].

The generation by the industry of wastewater with a high content of organic matter is considered a side-effect of some production processes. The treatment of such waters represents a cost, but also involves a loss of nutrients that are contained in these effluents or, when discharged without treatment to water bodies, they represent a very important source of ecological deterioration. On the other hand, agricultural production requires high volumes of water and fertilizer application to achieve optimal yields.

Therefore, the 'controlled' application on agricultural soils of industrial wastewater with a high content of organic matter intends to achieve the reuse of water and its contained nutrients. The goal is to reuse residual water that otherwise represents a cost and/or a potential environmental threat to the environment, and to recycle nutrients, incorporating the contained organic matter that will be transformed to nutrients that are required by agricultural crops.

It is important to mention that the application of wastewaters on agricultural soils is a practice carried out since ancient times. However, their use may affect the integrity of the soil and groundwater when the organic matter application is larger than the degradation/assimilation capacity of the soil. Large amounts of organic matter and water on the soil for long periods will cause depletion of oxygen; therefore, the anaerobic decomposition of the organic matter could cause the generation of methane, the reduction or loss of agricultural production, and the potential groundwater pollution with elements such as heavy metals, salts, etc.

This work presents a model developed to correlate factors and relationships between soil-plant-wastewater and to evaluate the implications of the quality of the wastewater on the soil and plants, depending on their properties and nutrient requirements/thresholds. To evaluate the model some calculated test cases are discussed.

The model is based on the application of a set of mathematical equations, taken from different authors, to estimate the optimal conditions for the application of wastewater to the soil. Equations with more conservative results were considered, in order to avoid saturation of the soil and groundwater pollution.

2. Application mechanisms of wastewaters on agricultural soils

We found three different practices for the application of wastewaters with high organic matter content. The first one is the *slow-rate application*, which consists of the application of a controlled hydraulic load on soil covered with determined vegetation. The soil, through percolation, filtrates the components of the wastewaters. The second practice is the *fast infiltration*, which is used for wastewaters that have received any type of pre-treatment and are applied in large amounts on highly permeable soils, allowing the waters to get quickly to the aquifer in the correct amount. Finally, *surface irrigation*, which implies the distribution of wastewaters on the surface of soils with vegetation coverage, controlled slopes, and low permeability. The objective of this treatment is the filtration of water through the runoff of the vegetation coverage [5].

This study uses the slow-rate application criteria, carried out intermittently to allow for ground aeration. The period between applications makes easier the degradation of organic matter and the nutrients of the wastewaters to bioavailable forms for the plants. Otherwise, the lack of oxygen causes an anaerobic decomposition that affects the development of the plant [6]. The slow-rate application can be of two types:

Type I, which applies the *maximum hydraulic load*. This type provides large amounts of water to small land surfaces and is mainly used in humid regions. This application criterion depends on the soil permeability or on the nitrogen discharge. Additionally deep percolation and soil transpiration should be considered [7].

Type II searches for the *irrigation optimal potential*. In this type, the water treatment is the secondary objective and the minimum amount of water to maintain the crops is applied, according to their water and nutrient requirements. This criterion is usually applied in dry lands. [7]

This study considers the two types described above for land application of the wastewater. The application of slow-rate Type I is a mechanism of organic matter removal from wastewater, while Type II is used to determine the minimum water volume to be applied in order to fulfil the physiological requirements of the crop. It is very important to consider that the amount of applied wastewater does not pretend to cover totally the crop's requirements; it only represents an additional amount of water that will benefit the crop's development.

Characteristics that the soil must fulfil for the slow-rate application (Type I and II) are described in Table 1.

The land application of wastewaters is to be based on a limiting design parameter, which controls the application design and determines the size of the hydraulic load rate that can be managed. In order to determine the limiting design parameter (LDP), the following aspects must be considered: hydraulic capacity of the soil, nitrogen contained in the wastewaters, biochemical oxygen demand (BOD_5) of the wastewaters, and the amount of toxic elements such as heavy metals [7]. The LPD is defined based on the organic matter content of the wastewater, the soil permeability, or the nitrogen concentration of the wastewater.

PARAMETERS	SLOW-RATE CHARACTERISTICS
Climatic Restrictions	Requires storage facilities during the rainy or cold season.
Water Table Depth	0.6–0.9 m
Slope	<20% in ploughed land <40% in non-ploughed land (trees)
Soil Permeability	0.5–50 cm/h
Permeability Interval	0.15–50.8 m/h
Soil Depth	> 0.61 m

Table 1. Soil characteristics required for slow-rate application. [5,7]

2.1. Wastewater-soil-crop relation

The relation existing between the wastewater, soil, and crops is mainly based in the equilibrium of the constituents of each one of these elements. In such a way, the vegetative cycle of the crop depends on the amount of water and nutrients available for its development. This availability is related to the humidity retention capacity of the soil and its fertility, which depends, among other factors, on the amount of organic matter present in the soil and provided by the crops at the end of their vegetative cycle. The decomposition of the organic vegetal material, under adequate conditions, allows the liberation of nutrients that contribute to soil fertility [8].

VARIABLES		
Wastewater	**Soil**	**Crop**
Biochemical Oxygen Demand	Texture	Nitrogen Requirement
Total Suspended Solids	Structure	Phosphorus Requirement
Fat and Oil	Permeability	Potassium Requirement
Metals	Water Table Depth	Toxicity to Metals
Nitrogen	Organic Matter	Sensibility to Salinity
Phosphorus	pH	Sensibility to Boron
Potassium		Evapotranspiration
Boron		
Salinity		
Sodium Absorption Relation		
Nitrogen-Carbon Relation		

Table 2. Characteristics of the variables in the soil-crop-water relation. [5,9–10]

The evaluation of wastewaters presented in this study is based on the relation existing between the water, the soil, and the crops, where the used water is an effluent of the alimentary and beverage industries. Therefore, the wastewaters present high contents of organic matter, suspended solids, and nutrients that modify these relations, affecting the soil, the crops, and the groundwater. The most relevant characteristics of the wastewaters, due to their potential effect on the previously described relations, are the concentration of Biochemical Oxygen Demand, total nitrogen, heavy metals, salinity, and boron.

The proper operation of the proposed application must control the main characteristics of each one of the variables of the relation, maintaining a balance between them. Table 2 lists the characteristics considered in the present application.

2.2. Removal mechanisms of the wastewater constituents

The present research uses the soil as the degradation/filtrating medium of the organic matter contained in the wastewater and the model is based on the diverse mechanisms, which are performed by the soil and the plant, for the removal of the constituents of the wastewater.

Bacteria, actinomycetes, and fungi, which are found in large amounts on the superficial layer of the soil, carry out the elimination of the BOD_5 from the applied wastewater. Once the organic matter has been degraded, some components become available as nutrients for the crops [7].

The removal mechanism of the BOD_5 depends on the application period, the drainage of wastewater, and the aeration period of the soil. The limits of the load that can be supplied to the soil must be based on the maintenance of the aerobic conditions of it, for which it is necessary to have an aeration period longer than the BOD_5 application time [11].

Consequently, the application must take into consideration that the BOD_5 of the applied organic matter is to be smaller than the amount of oxygen present in the soil.

The removal of nitrogen is complex due to the way this element is typically found in industrial wastewaters, such as: N_2, organic N, NH_3, NH_4, NO_2, and NO_3. [5] The manual for the design of soil wastewater treatment process of the US EPA [12] explains that the more oxidized the nitrogen form is, the more effective its retention and removal can be.

The nitrogen uptake of the crop is the main removal mechanism in the low-load application. However, denitrification and volatilization can be important depending on the wastewater constituents and the application place, according to the weather conditions.

There are different removal mechanisms of this element depending on the way it presents in the wastewater. The organic nitrogen can be trapped or filtrated by the soil; the ammonium (NH_4) and the ammonia (NH_3) can be volatilized, captured by the plant, or absorbed by the soil; and the nitrates (NO_3) can be absorbed by the plant or denitrified and liberated to the atmosphere in molecular nitrogen form [11].

The immobilization and storage of the nitrogen depends directly on the relation carbon/nitrogen (C/N), which determines the time in which the nitrogen is mineralized becoming available for the plant. Therefore, the optimum relation in the wastewaters for the application should be between 20:1 and 30:1 [13].

The metal removal of the wastewaters, according to [7], is given by the absorption of the soil, precipitation, ionic exchange, and complexation. The silt/clay or fine texture soils allow for almost complete removal of metals. In such a manner, plants have the capacity of removing metals from wastewaters by means of the evapotranspiration process that liberates the water from the environment and confines the metals in their structure. Consequently, these plants should not be used for feeding purposes. It is important to mention that for this research, the use of wastewaters with heavy metals is restricted.

The removal of phosphorus must take into account the fact that its presence in wastewaters is extensive. Even though risks to human health have not been reported, it is considered a risk to the quality of the water bodies due to its eutrophication potential. In the industrial wastewaters, phosphorus can be present as orthophosphate (PO_4)-3, polyphosphate, or organic phosphorus. Phosphates are immediately available for the use of the microorganisms existing in the soil but they are not present in the plant [6].

In the soil, the removal mechanism of the phosphorus depends on the chemical relations carried out through long periods; therefore, the continuous discharge of wastewaters with phosphorus will reduce the retention capacity of this element. However, according to [9], this capacity will not be exhausted. On the other hand, the phosphorus removal due to the plant intake is achieved in 20-25% of cases. In this case, it is necessary for the soils to contain iron and aluminium oxides and calcareous substances so the removal of this element increases proportionally to the clay content and the time of contact with the soil.

Sodium removal from the wastewaters is caused by the cationic exchange in the particles of the soil and is due to the crop's absorption of salts, which accumulates these elements in its structure, depending on its tolerance to salinity [10]. Sodium is an element considered in this research because its frequent presence at a high-level in the wastewaters may cause a detrimental effect on the crops due to the salinity or alkalinity that it generates. The excess of sodium with respect to the magnesium and the calcium has an effect over the osmotic potential, which reduces the water absorbing capacity of the plant. It also affects the soil structure, preventing the clay from retaining the cations therefore modifying the hydraulic capacity of the soil profile [14].

Boron removal is carried out by the soil as long as iron and aluminium oxides are present in its composition. Boron is an essential micronutrient for plants, but it can be toxic at relatively low concentrations. For the application of wastewaters, it can be assumed that all the applied boron that is not assimilated by the plants will be leached to the groundwater. Therefore, this element is to be considered in the monitoring process (daily crop inspections) of the application, in order to prevent contamination of water bodies and plant toxicity [9].

2.3. Characteristics of the soil and its role in the application of wastewaters

The physical, chemical, and biological characteristics of the soil determine the behaviour of the water in the soil. Therefore, in this part of the document, each of the characteristics of the soil will be described, as well as their influence on the application of wastewaters.

The texture refers to the size of the particles that form the soil, such as: clay, silt, and clay-silt. In the context of the application of wastewaters with a high content of organic matter, the size of the particle will influence the filtration and percolation capacity of the soil. Consequently, fine textures present a slow-infiltration and percolation, optimal for superficial irrigation; medium textures adapt better to low load rate application; and coarse textures allow for the application of the process of fast infiltration [6].

The structure represents the shape and degree of the particle aggregation, determining the water and air movement, and the porosity. The soil aggregates form pores, which allow for the conduction of air and water that defines the infiltration capacity of the soil [6].

The soil depth allows for the retention of the water particles depending on the time of their presence in the soil. This time depends on the application rate and the permeability of the soil. On the other hand, an adequate depth of soil allows the development of roots, microbial activity, and the separation between the wastewater and the saturated area [15].

The chemical properties of the soil influence the growth of the plant due to the nutrients' availability, the purification of wastewaters, and the hydraulic conductivity. The pH of the soil is a key variable that affects the physical, chemical, and biological properties of the soil. It is affected by different factors such as the precipitation rate, the irrigation water quality, the dissociation of the carbonic acid, the organic matter content, the weathering of minerals, the presence of polymers and aluminium hydroxide, and the application of nitrogenized fertilizers. Additionally, the pH of the soil has an influence on the solubility of different compounds, the activity of microorganisms in the soil, the crop's growth given the availability of nutrients and metals, the mobility of the chemical constituents of the soil, the clay dispersion, and the formation of aggregates. The application of wastewaters with a low pH for long periods may affect the fertility of the soil and allow the leaching of metals. Most of the crops are properly developed in a pH range between 5.5 and 7.0 [16].

The buffering capacity of the soil is related to the amount of organic matter that is contained. This property prevents drastic fluctuations of the pH that may affect the plant and the microorganisms and, additionally, favours the capacity of cationic exchange [9].

The content of organic matter in the soil has an influence on the structure and provides energy to the microbial activity that allows for the formation of aggregates. A large amount of organic matter supposes a better structure and therefore a better water retention. On the other hand, the decomposition of organic matter forms humic substances that react with the clay particles (silicates), iron, and aluminium oxides forming bonds among the soil particles. This characteristic favours the capacity of cationic exchange since the nutrients are retained for the crops, as well as the metallic cations and the organic chemicals, due to the larger specific surface of the soil particles [16].

The amount of organic matter is related to the absorption and availability of nutrients (micronutrients) for the plant, allowing the formation and availability of stable compounds with polyvalent cations — such as Fe_3+, Cu_2+, Ca_2+, Mn_2+, y Zn_2+ — reduces the metallic capitation by the crops and their mobility in the soil [9].

2.4. Objective of the crop in the application of wastewater

The role of the plant in wastewater application is to absorb the majority of the nutrients, depending on the type of crop, that are applied in order to convert them into biomass [12]. In this way, crops with larger nutritional requirements will extract larger amounts of nutrients, allowing the removal of larger amounts of them from the applied wastewater and to fitoremediate some contaminants, depending on associated bacteria at the roots.

In the application of wastewater at a low load rate, the role of the plants is mainly the removal of the nitrogen and in some cases the generation of a financial benefit by means of crop fertilization, preventing erosion, and increasing the infiltration rate. Usually, the wastewater does not have an appropriate C/N relation of the range between 20:1 and 30:1. However, the roots of the plants provide a source of organic carbon that helps with the process [6].

The evapotranspiration is the process that determines the amount of water that the crop requires for its physiological processes, considering the plant transpiration and the evaporation from the plant and ground surfaces. The amount of evapotranspiration depends on the atmospheric conditions: such as the solar radiation, air temperature, relative humidity, and the wind speed, as well as the water availability from the soil [11].

In summary, the present work is based on the interactions existing between the wastewater, soil, and crop — where the wastewater provides the nutrients in organic form during the percolation; the soil is the medium that is in charge of the removal, degradation, and storage of such nutrients; and the crop performs the extraction of the nutrients and their transformation into biomass. At the end of the process, the organic constituents of the percolated water are expected to be reduced, avoiding contamination of groundwater.

The success in the application of effluents with a high organic load onto the soil will depend on the correct interpretation of the phenomena that occur in the soil and its relation with the plant. For this reason, the computer model STAR ASA has been developed. This model uses a set of mathematical equations taken from different authors, that, when applied within a process, allow the estimation of the conditions to be considered when the wastewaters are applied to the soil, making the decision process easier. The selected models were selected from those that obtained positive results and the most conservative results, in order to avoid the saturation of the soil and the contamination of groundwater. The application of this model will provide the wastewater volume that can be applied to the soil, the required land surface for the application, the minimum necessary time between applications, and the amount of nutrients incorporated to the soil.

3. Selected mathematical models

The characterization of the wastewater presented in this study is based on the application of the STAR ASA model.

3.1. Water balance

In the low load application Type I, the water balance allows for the determination of the hydraulic load that has to be applied to the soil to at least fulfil the crop's requirement, considering the part of water that percolates to the groundwater based on the permeability of the least permeable layer. Equation 1 shows the balance between the applied water (depth and precipitation) and use of water due to percolation and evapotranspiration [5, 9, 11]. Figure 1 shows a conceptual framework of this balance.

$$L_w = Et_c - Pr + P_w \tag{1}$$

L_w= Wastewater hydraulic loading rate — based on soil permeability [mm/d]

Et_c = Evapotranspiration rate [mm/d]

Pr = Precipitation rate [mm/d]

P_w =Design percolation rate [mm/d]

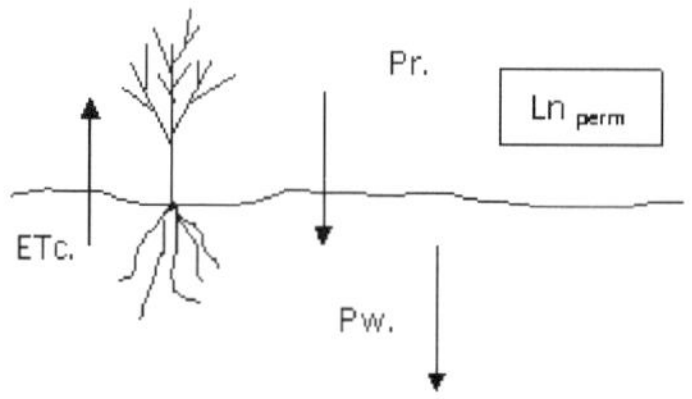

Figure 1. Water balance framework based on the soil permeability

L_n= Wastewater hydraulic loading rate — based on soil permeability [mm/d]

Et_c = Evapotranspiration rate [mm/d]

Pr = Precipitation rate [mm/d]

P_w =Design percolation rate [mm/d]

In the application of low load Type II, the water balance focuses on the better use of the irrigation water than in a water treatment system. However, it is considered in this study since it allows the determination of the minimum hydraulic load rate required by the crop for its optimal development. This model includes in the calculation the leaching requirement and the irrigation efficiency, which is defined as the capacity of irrigating a determined volume of water into a uniform land surface. Equation 2 [6,9] shows this balance, which can also be seen in the conceptual framework of Figure 2.

$$Lw = \left(Et_c - Pr\right) * \frac{\left(1 + LR\right)}{E_i} \tag{2}$$

Lw = Wastewater hydraulic loading [mm/d]

Et_c = Crop evapotranspiration [mm/d]

Pr = Precipitation rate [mm/d]

LR = Leaching requirement (fraction)

E_i = Distribution system efficiency (fraction)

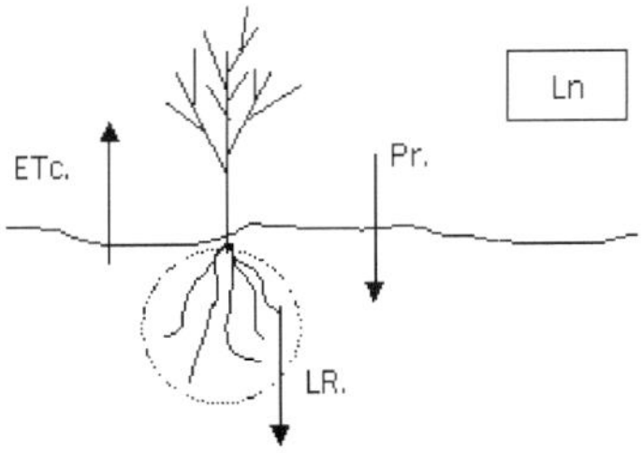

Figure 2. Water balance framework based on irrigation

The leaching requirement is used in the application of low load Type II in order to estimate the amount of water to be added in addition to the crop's requirement, avoiding the accumulation of salts in the root area of the plant, therefore preventing adverse effects for the crop development. Equation 3 [10, 14] represents this calculation, with help of the electric conductivity of the applied wastewater and the electric conductivity that the soil can withstand without affecting its performance.

$$LR = \frac{CE_w}{5(CE_e) - CE_w} \quad (3)$$

LR = Leaching requirement (fraction)

CE_w = Average conductivity of irrigation water [dS/m]

CE_e = Required conductivity in drainage water to protect de crop [dS/m]

L_n= Wastewater hydraulic loading rate — based on soil permeability [mm/d]

Et_c = Evapotranspiration rate [mm/d]

Pr = Precipitation rate [mm/d]

LR = Leaching requirement (fraction)

3.2. Oxygen balance

As mentioned previously, the application of wastewater from the food and beverage industries implies the use of large concentrations of organic matter with a high BOD_5. Consequently, it is important to estimate the amount of equivalent oxygen required for the oxidation of this organic matter by means of the microorganisms, maintaining aerobic conditions in the soil. It must be considered that the BOD_5 concentration does not show the Total Oxygen Demand (TOD) existing in the wastewaters. Therefore, Equation 4 [12] estimates the TOD by means of the addition of the BOD_5 and the concentration of the oxygen demand of the nitrogenous compounds (NOD):

$$TOD = BOD + NOD \quad (4)$$

TOD = Total oxygen demand [g/m3 ó mg/l]

BOD_5 = Biochemical oxygen demand [g/m3 ó mg/l]

NOD = Nitrogenous oxygen demand [g/m3 ó mg/l]

$$NOD = 4.56 * \text{Nitrifiable Nitrogen} \quad (5)$$

The oxygen demand by the nitrogenized compounds is calculated from the wastewater nitrificable nitrogen concentration [mg N-NH3l] multiplied by an estequiometric coefficient equal to 4.56 mg of oxygen [12, 17].

3.3. Analysis of the diffusive transportation in the soil

In this study, the estimate of the soil oxygenation by means of the application of wastewaters with a load of organic matter is performed considering the diffusive transportation of the oxygen through the soil as the main source of oxygenation. To prove this supposition, it was necessary to use the analytic solution developed by Papendrick and Runkles [18], based on the second law of Fick (Equation 6). This solution allows the estimation of the concentration of oxygen in the soil, at a given and determined depth and time, considering a constant respiration (Equation 7). The solution proposed by these authors was proven by Kanwar [19] and Prasanta [20], demonstrating that the flux of oxygen (mass O_2/area x time) is directly proportional to the gradient of oxygen concentration between the atmosphere and the soil, and with the microbial oxidation rate. Those researchers consider the microbial oxidation rate to be constant. However, in field conditions, the respiration rate increases with the fertility of the soil and consequently the flux of oxygen in the soil.

In this study, the estimate of the oxygenation of the land irrigated with wastewater with high loads of organic matter, considers the oxygen diffusive transport through the soil as the main source of oxygen, which is used by the model developed in [18] that calculates the concentration of oxygen in depth and time.

To calculate the required oxygen flow for soil aeration, an equation is used that considers the elapsed time, the diffusivity of oxygen in soil and the oxygen concentration in the atmosphere and soil [12, 17], through a balance between the amount of equivalent oxygen that requires the organic matter from wastewater and the time that must elapse to reach the oxygenation of the soil through diffusive transport.

$$\frac{\partial C}{\partial t} = D\frac{\partial^2 C}{\partial x^2} - \alpha \tag{6}$$

The initial and threshold conditions for the application of this model are:

$$C(x, 0) = Ct$$
$$C(0, t) = Co$$
$$\delta c(,t) / \delta x = 0$$

Where:

C(x, t) = Oxygen concentration at a given depth (x) and at a time (t) [cm3 O2/cm3 air]

Ct = Initial concentration of oxygen in the ground [cm3 O2 / cm3 air]

Co = Concentration of oxygen in the atmosphere [cm3 O2 / cm3 air]

t = Time [h]

D = Diffusivity of oxygen in the soil [cm2 / h]

x = Depth [cm]

α = Edaphic respiration rate [h-1]

The following solution is established:

$$C(x,t) = C_t - \alpha t + (C_o - C_t)*\mathrm{erfc}\frac{x}{2\sqrt{D*t}} + \alpha\left[\left(t + \frac{x^2}{2D}\right)\mathrm{erfc}\frac{x}{2\sqrt{D*t}} - x\sqrt{\left(\frac{t}{\pi * D}\right)}*e^{\frac{-(x^2)}{4D*t}}\right] \quad (7)$$

Where:

erfc = Complementary error function

In order to determine the oxygen flux in the soil, Equation 7 was applied with typical values of oxygen respiration and diffusivity (D). Table 4 shows three sets of values or tests used in the research.

	1st Test (P1)	2ª Test (P2)	3ª Test (P3)
Initial concentration of oxygen (Ct) (cm3 of O2/cm3 of air in the soil)	0.21	0.21	0.21
Respiration (α) (g of O2/m3 of air in the soil*d)	0.002125	0.0025	0.0125
Diffusivity of oxygen in the soil (D) (cm2 of soil / h)	259.2	89.1	89.1

Table 3. Typical values of oxygen respiration and diffusivity.

In each test, an equation with seven depth values (x) and six different aeration times (t) was applied. The relative concentrations of oxygen obtained in each case are shown in Table 4.

		Relative Concentration of Oxygen (C/C0)					
	X (cm)	t = 0	t = 4	t = 8	t = 12	t = 16	t = 20
P1	0	1	1	1	1	1	1
	15	1	0.983	0.974	0.9673	0.20997	0.9566
	30	1	0.972	0.955	0.9421	0.20995	0.9210
	45	1	0.966	0.942	0.9231	0.20994	0.8922
	60	1	0.962	0.933	0.9090	0.20994	0.8691
	75	1	0.961	0.927	0.8990	0.20993	0.8510
	90	1	0.960	0.924	0.8919	0.20993	0.8369
P2	0	1	1	1	1	1	1
	15	1	0.970	0.953	0.9398	0.92853	0.9186
	30	1	0.958	0.927	0.9021	0.88051	0.8613
	45	1	0.954	0.914	0.8800	0.84983	0.8225

http://dx.doi.org/10.5772/59999

	Relative Concentration of Oxygen (C/C0)						
	X (cm)	t = 0	t = 4	t = 8	t = 12	t = 16	t = 20
	60	1	0.953	0.908	0.8680	0.83124	0.7974
	75	1	0.952	0.906	0.8619	0.82060	0.7818
	90	1	0.952	0.905	0.8591	0.81486	0.7726
P3	0	1	1	1	1	1	1
	15	1	0.851	0.765	0.6990	0.64263	0.5928
	30	1	0.789	0.635	0.5103	0.40253	0.3063
	45	1	0.769	0.570	0.3999	0.24913	0.1125
	60	1	0.763	0.541	0.3398	0.15621	0
	75	1	0.762	0.530	0.3095	0.10302	0
	90	1	0.762	0.525	0.2954	0.07432	0

Table 4. Oxygen concentrations with values of α and D.

Once the relative concentrations of oxygen are calculated, the curves were drawn for each test, obtaining the polynomial equations based on the depth, as shown in Figures 3, 4, and 5.

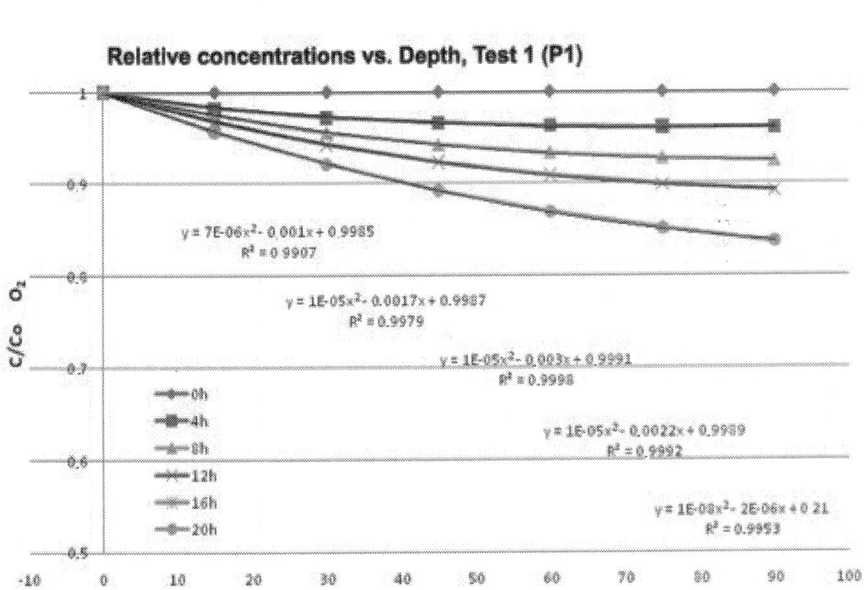

Figure 3. Relative concentrations vs. depth, Test 1 (P1).

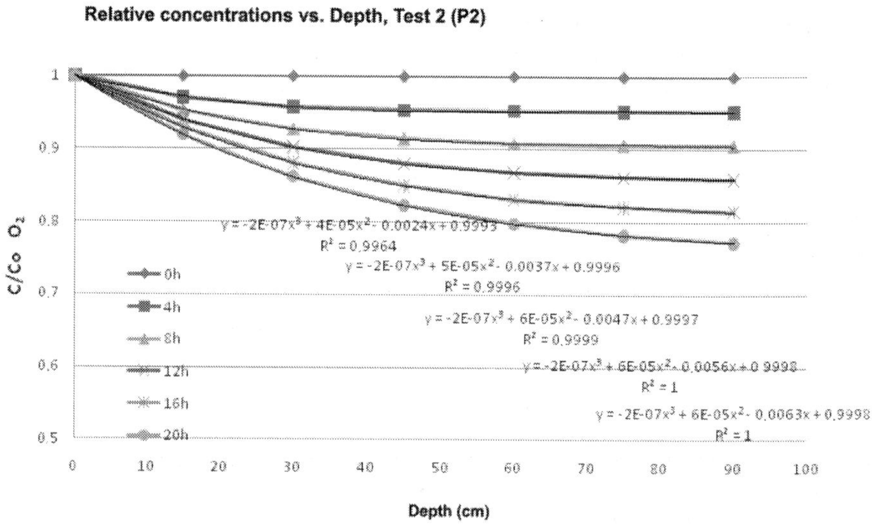

Figure 4. Relative concentrations vs. depth, Test 2 (P2).

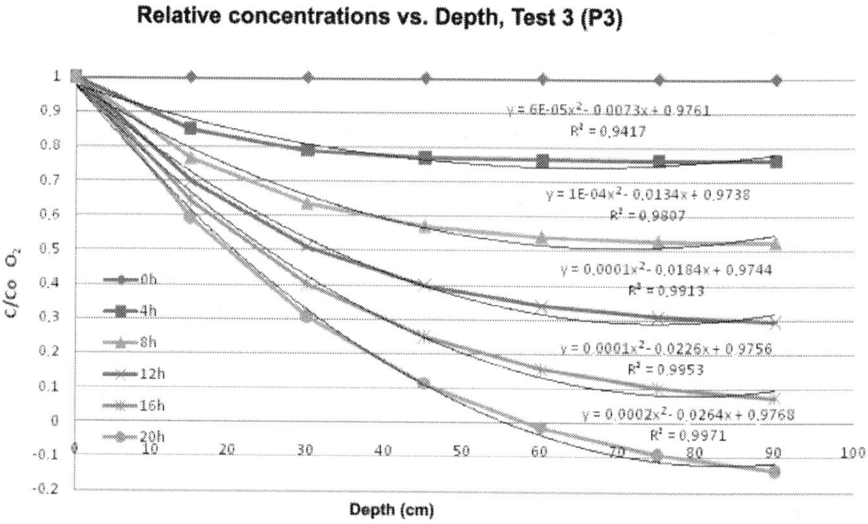

Figure 5. Relative concentrations vs. depth, Test 3 (P3).

Using the previous diagrams, the polynomial equations that describe the profile of the oxygen concentration in the soil for each time step were obtained. The derivation of the polynomial, evaluated at x=0 (soil surface), presents the flux for each time. For a constant degradation, in a given period, the area under the curve flux vs. time will calculate the amount of consumed oxygen. In this way, the mass of oxygen represents the load of BOD that could be accepted per day. Table 6 shows the oxygen flux in the soil for each one of the performed tests.

Test 1		Test 2		Test 3	
9.6	cm^3/cm^2	7.18	cm^3/cm^2	27.068	cm^3/cm^2
0.0096	L/cm^2	0.00718	L/cm^2	0.027068	L/cm^2
0.0004	mol/cm^2	0.00032036	mol/cm^2	0.00120775	mol/cm^2
136.5	g/m^2	102.5	g/m^2	386.5	g/m^2

Table 5. Oxygen flux estimated using typical values of respiration and diffusivity.

With these results, the flux of oxygen that the soil can receive, using typical values of respiration and diffusivity, was determined to be between 102.5 y 386.5 g/m^2. On the other hand, it was observed that the flux of oxygen in the soil is directly proportional to the increase of the edaphic respiration and depends on the coefficient of diffusivity of the soil. These two factors determine the increase of the gradient of the oxygen concentration of the soil, which increases the flux of oxygen.

Considering the obtained results, this research established the application of wastewater with oxygen requirements lower than 100g/m^2 as a maximum, avoiding soil anaerobic conditions.

3.4. Estimation of oxygen flux due to diffusive transportation

Once it is established that the aeration of the soil is produced mainly by diffusive transportation phenomena, the estimation of the oxygen flux using Equation 7 will require information of the edaphic respiration, which is only obtained by field experimentation; this constitutes a limitation for the application of this model.

For this reason, the estimation of the oxygen flux in this research will be performed considering a simplified diffusive model that will require no information related to the edaphic respiration but will allow the adequate estimation of a minimum level of oxygen flux in the soil.

$$\frac{\partial C}{\partial t} = D\frac{\partial^2 C}{\partial x^2} \tag{8}$$

In order to determine the oxygen flux in a simplified manner, this research selected the analytic solution of the Law of Fick (Equation 8), proposed in [17] — the determination of the oxygen flux considers the elapsed time, the oxygen diffusivity of the soil, and the concentration of oxygen in the atmosphere and air in the soil, but it does not consider the edaphic respiration, as shown in Equation 9. This model establishes a minimum concentration of oxygen in the root area of the soil in order to avoid negative effects in the crop development.

$$N_{O2} = 2(C_{O2} - C_P)*\sqrt{Dp*t_a/\pi} \quad (9)$$

N_{O2} = Flux of oxygen crossing the soil surface [g/m2]

C_{O2} = Vapour phase O2 concentration above the soil surface [g/m^3 o mg/l]

Cp = Vapour phase O2 concentration required in soil to prevent adverse yields or root growth [g/m^3 o mg/l]

Dp = Effective diffusion coefficient [m2/d]

t_a = Aeration time

$$Dp = 0.66*\ \epsilon*S*Do2 \quad (10)$$

Tortuosity = 0.66 (fraction)

ϵ= Porosity (fraction)

S = Fraction of air filled soil pore volume at field capacity (fraction)

D_{O2} = Oxygen diffusivity in air [m2/d]

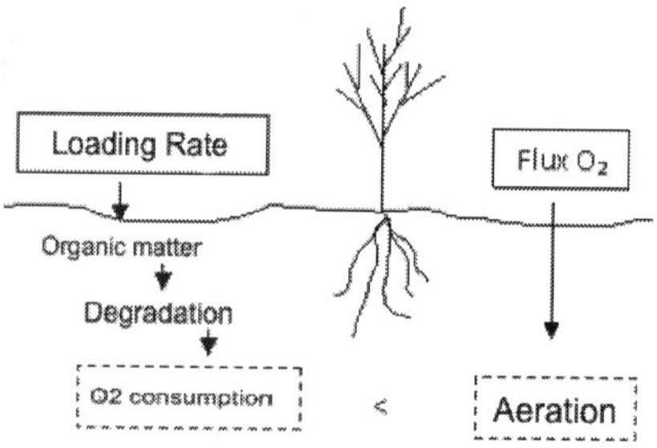

Figure 6. Outline of oxygen balance

The adequate flux of soil oxygen is achieved by means of a balance between the amount of equivalent oxygen required by the organic matter present in the wastewater (Equation 4) and the time period needed in order to achieve the soil oxygenation by diffusive transportation, expressed in Equation 9. This balance can be observed in the framework shown in Figure 6.

http://dx.doi.org/10.5772/59999

3.5. Time between applications

In order to determine the time that is needed between applications of wastewater, Equation 10 is used [12]. This equation considers, in addition to the aeration time, the time that it takes to infiltrate the water in the soil (Equation 11) and the duration of irrigation.

$$t_{ap} = t_{air} + t_{in} + t_r \quad (11)$$

t_{ap} = Time between applications[d]

t_{air} = Time of aeration [d]

t_{in} = Time of infiltration [d]

t_r = Time of irrigation [d]

$$t_{in} = \frac{Ln}{k} \quad (12)$$

Ln = Hydraulic load rate (depth) [mm]

k = Infiltration or permeability in the saturated soil [mm/d]

3.6. Nitrogen balance

This balance is established through the determination of the hydraulic rate of the wastewater to be applied based on its nitrogen content; the following process differs from the design of the hydraulic load developed in Equations 1 and 2. In this case, the required hydraulic load is estimated based on the amount of nitrogen that can be removed by a crop as part of its nutritional requirements, in a determined land area and time. Equation 13 determines the nitrogen load that is to be incorporated into the soil for the crop to remove it adequately. This equation also considers the factor of nitrogen loss due to denitrification and volatilization. The result of this equation, divided by the nitrogen concentration in the wastewater, determines the hydraulic load rate as shown in Equation 14 [12].

$$L_n = \frac{U}{(1-f)} \quad (13)$$

L_n = Nitrogen loading [kg/(ha*year)]

U = Estimated crop uptake as a function of yield [kg/(ha*year)]

f = Nitrogen loss factor (0.5–0.8) (fraction)

$$Ln_{nitrogen} = \frac{L_{nitrogen}}{C_{nitrogen}} {}^{*}F \quad (14)$$

$L_{n\ nitrogen}$ = Hydraulic nitrogen load based on the nitrogen load [mm/d]

$C_{nitrogen}$ = Nitrogen concentration in the waste water [mg/l or g/m3]

F = Conversion factor

3.7. Application area

The determination of the required area for the application of wastewater in the soil is obtained from the relation between the volume of wastewater discharge and the hydraulic load rate to be applied in the soil, as shown in Equation 14 [6, 9, 12].

$$A = \frac{Q}{Ln_d} * F \tag{15}$$

A = Area required for the wastewater application [ha]

Q = Wastewater discharge volume [m3/d]

L_{nd} = Design hydraulic load rate [mm/d]

F = Conversion factor

4. Case of study

In order to understand the model, a set of tests were carried out based on nine (9) potential cases, considering the following conditions:

1. The selected crop in all the cases was sugar cane.
2. For all the cases, the wastewater is used to irrigate soils with agricultural potential: Deep soils (1.5 m to 3 m), with slopes less than 15° and with a loamy texture.
3. The climate conditions are similar for all cases. The considered average precipitation is 730 mm/year.
4. The most important constituents in the wastewater to be used are: Biochemical Oxygen Demand (BOD_5), Chemical Oxygen Demand (COD), and Total Suspended Solids (TSS). For these tests, DBO_5 values of 1,000, 25,000 and 50,000 mg/l were evaluated.

The use of the model for the decision-making process begins with the input of the information related to the area selected in order to determine the site's aptitude. The aptitude is valued in a general/ global manner considering some variables such as the slope and the depth of soil and groundwater.

The software requires the input of the concentration of the constituents corresponding to the wastewater to be applied (see Table 6). Three BOD_5 concentrations (1,000, 25,000, and 50,000 mg/l) were used for the tests. Those concentrations are frequently found in the food and beverage industry effluents.

http://dx.doi.org/10.5772/59999

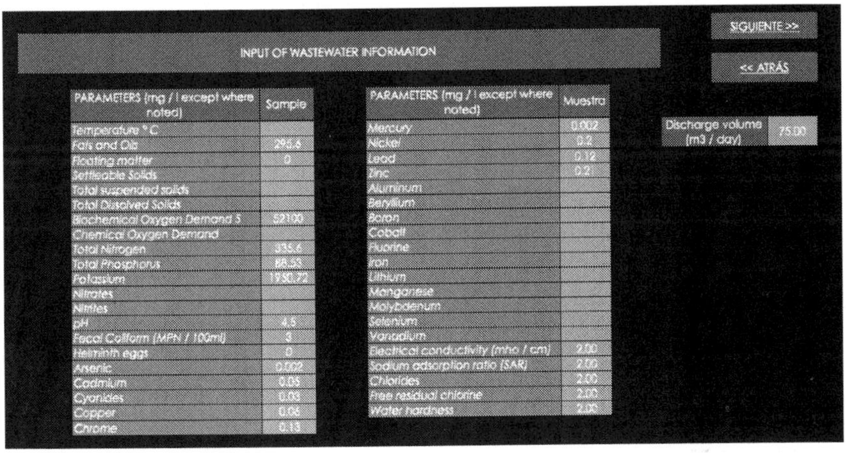

Figure 7. Input of wastewater information

Next, the software requires the input of information related to the characteristics of the soil (Figure 8). It is important to mention that the tests were performed considering loam, clay loam, and sandy loam textures for the soil. The diameter of the particles varied between 0.05 and 0.002 mm. These characteristics are related to other physical properties of the soil such as porosity and infiltration.

Additionally, the model requires information related to the type of crop present and the climatic conditions of the environment. For the performance tests, sugar cane crops with average precipitation conditions (730mm/year) were considered.

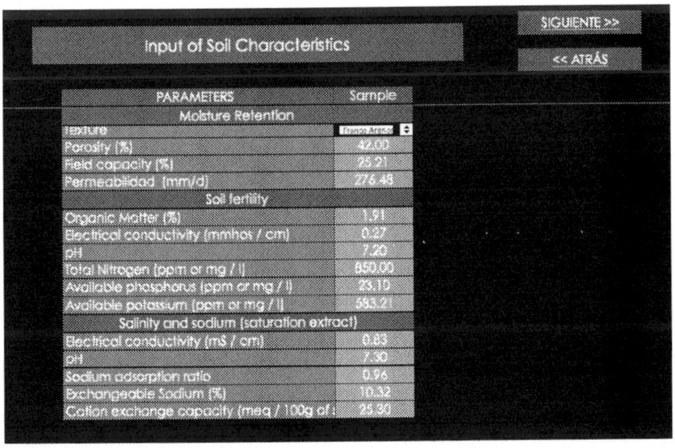

Figure 8. Input of soil characteristics

Once the information input is completed, the model calculates the optimum parameters for the wastewater application based on the available information. Figure 9 shows an example of the obtained results.

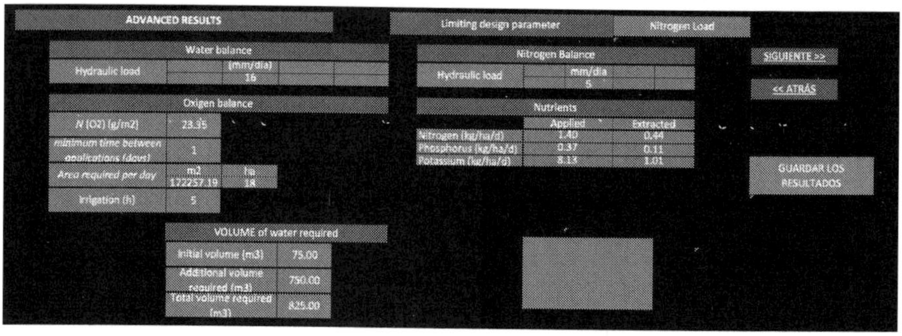

Figure 9. Advanced results

The results obtained from all performed tests in the hypothetical cases are detailed in Table 6. Those results show the model estimates of the required surface for the application and the estimated time lap between applications in order to achieve an adequate use of the nutrients provided by the wastewater, which will be required for the crop development.

TEST	BOD [mg/l]	Texture	Surface (Ha)	Time (h)
1	1000	Loam	0.11	6.48
2	1000	Sandy Loam	0.14	5.76
3	1000	Sandy Loam	0.10	11.28
4	25000	Loam	4.03	22.32
5	25000	Clay Loam	4.63	16.8
6	25000	Clay Loam	4.66	46.56
7	50000	Loam	9.18	50.64
8	50000	Clay Loam	7.91	28.8
9	50000	Clay Loam	14.89	149.04

Table 6. Model tests results

The analysis of the results of the tests from the STAR ASA model could determine that the higher the concentration of organic matter in the wastewater, the larger the requirement of agricultural area. However, this quantity depends strongly on the characteristics of the soil. The smaller the soil particle size, the smaller the required area (Figure 10).

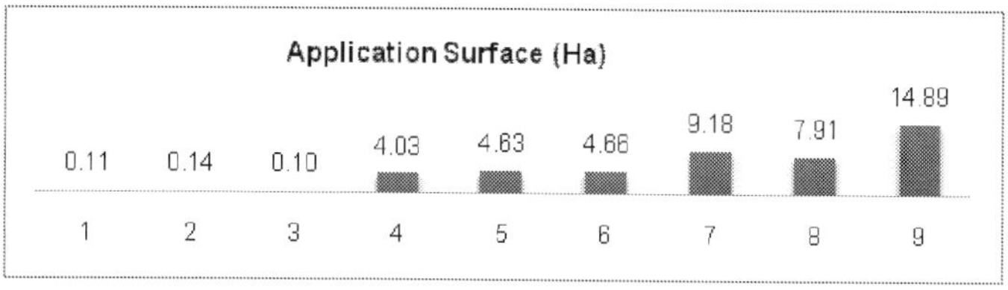

Figure 10. Results — application surface

The time required between wastewater applications depends mostly on the soil's characteristics. The smaller the particle size, the shorter the required time lap (Figure 11). This result is related to the infiltration capacity of the soils. The larger the particle size, the smaller the water retention period in the agricultural areas. Therefore, frequent application on sandy soils can cause the lixiviation of the nutrients and chemicals into the groundwater.

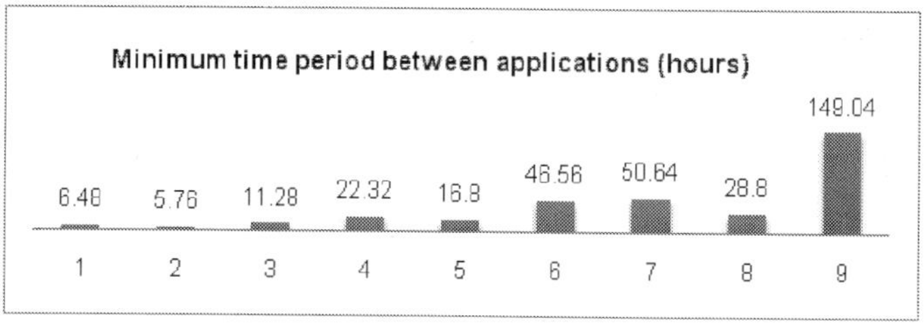

Figure 11. Results — time period between applications

5. Conclusions

The STAR ASA model was able to integrate the characteristics of the wastewater and the soil, as well as the nutritional requirements of the crops and the environmental regulations, by means of mathematical models selected to evaluate the feasibility of the application of wastewaters with a high content of organic matter with agricultural soil irrigation. The model allowed the estimation of the hydraulic load rate, the time lap between applications, and the amount of nutrients provided to the soil.

Based on the theoretical analysis of the used models, it was determined that the controlled application of wastewater allows the filtration and degradation of the constituents with a high content of organic matter that were applied to the soil, as long as the parameters recommended by the STAR ASA model are fulfilled.

Finally, considering the characteristics of the analysed wastewaters, the characteristics of the selected soils, and the nutritional requirements of the sugar cane, it was determined that for the test cases, the application of the effluents in the soil will contribute to fertilization, specifically nitrogen-related fertilization. However, it needs to be clarified that the results obtained in this study cannot be generalized to other type of effluent without previous analysis. The wastewaters, the agricultural soils, the climatic conditions, and the crops requirements may be different.

6. Recommendations

Considering this research as starting point, it is recommended to apply the model in additional case studies, for industries with wastewaters with different concentrations and constituents, in order to determine the behaviour of the application on those conditions to carry out the necessary modifications.

It is recommended to perform a study to determine the effect of the long-term application of organic matter versus the microbial respiration, as a complement for this research.

Monitoring is of vital importance for the correct application of effluents; consequently, it is recommended to perform control measurements at least every six months.

It is recommended to carry out an adequate dimensioning of the storage system of wastewater in containers considering the climatic conditions and the crop type, in order to avoid ponding, runoffs, leaching, and erosion in high precipitation and harvesting seasons.

Author details

Esteban Jácome Sandoval and Martin H. Bremer- Bremer*

*Address all correspondence to: mbremer@itesm.mx

Instituto Tecnológico y de Estudios Superiores de Monterrey, Mexico

References

[1] United Nations World Water Assessment Programme. Water in a Changing World. In U. Nations, *World Water Development* (pag. Report 3). Paris: UNESCO; 2009

[2] Lanz K., Müller L., Rentsch C., and Schwarzenbach R. De Quién es el Agua. Madrid: Gustavo Gili; 2008

[3] United Nations World Water Assessment Programme. Water and Energy. In U. Nations, World Water Development Report (pag. 2). Paris: UNESCO; 2014

[4] Siebert S., Henrich V., Burke J., and Frenken K. Update of the Digital Global Map of Irrigation Areas to Version 5. Germany: Institute of Crop Science and Resource Conservation — Rheinische Friedrich-Wilhelms-Universität Bonn; 2013

[5] Metcalf and Eddy. Water Reuse: Issues, Technologies and Applications. New York: McGraw-Hill; 2007

[6] Romero J. Tratamiento de Aguas Residuales: Teoría y Principios de Diseño. Bogotá: Editorial Escuela Colombiana de Ingeniería; 2004

[7] Crites R. and Tchobanoglous G. Sistemas de Manejo de Aguas Residuales, para Núcleos Pequeños y Descentralizados (Vol. II). Bogotá: McGraw-Hill Interamericana; 2000

[8] Asawa, G. Irrigation and Water Resources Engineering. New Delhi: New Age International (P) Limited Publishers; 2005

[9] Crites R., Reed S., and Bastian R. Land Treatment Systems for Municipal and Industrial Wastes. USA: McGraw-Hill; 2006

[10] Lazarova V. and Bahri A. Water Reuse for Irrigation. Boca Ratón: CRC Press; 2005

[11] Reed S., Crites R., and MiddleBrooks EJ. Natural Systems for Waste Management and Tratment (2nd Edition ed.). USA: McGraw-Hill; 1995

[12] US EPA. Process Design Manual: Land Treatment of Municipal Waste Effluents. Cincinnati: US Environmental Protection Agency; 2006

[13] Urbano P. *Fitotecnia: Ingeniería de la Producción Vegetal.* Madrid: Mundi-Prensa; 2002

[14] Ayers R. and Westcot, D. Water Quality for Agriculture. California: FAO; 1994

[15] Delleur J., ed. Handbook of Groundwater Engineering. Boca Raton: CRC Press; 2007

[16] Navarro G. Química Agrícola. Madrid: Mundi-Prensa; 2000

[17] McMichael F. and McKee J. Wastewater Reclamation at Whittier Narrows. Pasadena: Environmental Health Engineering, California Institute of Technology; 1965

[18] Papendick R. and Runkles J. Transient-state Oxygen Diffusion in Soil: I. The Case when Rate of Oxygen Consumption is Constant. Soil Science; 1965, p. 251–261.

[19] Kanwar R. Analytical Solutions of the Transient State Oxygen Diffusion Equation in Soils with a Production Term. J. Agronomy and Crop Science; 1986, p. 101–109.

[20] Prasanta K. Transient Finite Element Method Solution of Oxygen Diffusion in Soil. Ecological Modelling; 1999, p. 227–236.

Chapter 3

Wastes in Building Materials Industry

Marinela Barbuta, Roxana Dana Bucur, Sorin Mihai Cimpeanu, Gigel Paraschiv and Daniel Bucur

Additional information is available at the end of the chapter

http://dx.doi.org/10.5772/59933

1. Introduction

In the last decades, due to the modern lifestyle, the progresses in industry and technology had led to an important increase in the amount and type of wastes. The problem of waste accumulation every year is all over the world. These industrial and agricultural wastes are by-products, slag, rice husk ash, bagasse, fly ash, cement dust, brick dust, sludge, glass, tires, etc. The wastes represent a major problem for the environment because the air pollution (the dust and very fine particles which spread in the atmosphere) and leaching toxic chemicals (arsenic, beryllium, boron, cadmium, cobalt, lead, manganese, mercury, molybdenum, selenium, strontium, thallium, hydrocarbon compounds, etc.) when are dumped in landfills, quarries, rivers or oceans. The capitalization of waste is difficult because of their variety, as well as their unknown properties over time.

Lately, the environmental sustainability became an important problem from the point of view of natural resources and that of wastes. The construction and the building materials sectors are involved in both processes: building industry is the largest user of natural materials and in addition a large amount of wastes results from the demolition of constructions.

The building material industry is a domain of interest for using the wastes and researchers have tried to produce new construction materials incorporating wastes. The new generation of building materials is developing on other theories in concordance with the sustainability of environment.

Concrete is obtained from natural aggregates, cement and water, compounds which make it a cheap material and easy to produce anywhere. Usually, ordinary concrete contains about 12% cement, 8% water and 80% aggregates by mass. Aggregates and water are from natural resources, only cement must be produced in fabrics, processes which are polluted the environment (for producing 1 m^3 of concrete a quantity of 480 kg of CO_2 is liberated in the atmosphere). For reducing the aggregate and cement consumption, the replacing materials obtained from wastes were studied.

Because the cement industry is responsible for 5-7% of worldwide emission of CO_2, (which means 1.6 billion tons of carbon dioxide into the atmosphere), in preparing concrete, the cement dosage can be reduced by using mineral additions, strategy that also can contribute to environment protection by preserving the energy and consume the huge quantities of wastes.

Near cement concrete other building materials are obtained by using wastes, such as: high strength concrete, which has in the mix different additions (silica fume, fly ash, etc.), asphalt concrete, bricks, pavements, roof tiles, prefabricated units, claddings, etc.

Some building materials are totally obtained from wastes, such as "green" materials. The new concept of green buildings offers more energy-and resource efficiency. This concept means the greening of building industry by using only green materials. The technologies of obtaining green materials are available, but their use in construction industry is limited.

The building material and construction industry is one of the principal users of wastes in the processes for obtaining materials or products, for constructing bridges or highways, in soil stabilization, in hydraulic construction, etc. From environmental considerations an extensive waste utilization in construction is recommended, although particular wastes may be too risky to use.

2. Waste classification

For many years a lot of wastes have been accumulated in the entire world and they influenced the environment and people life. The necessity of eliminating or at least, reduction of huge quantities of wastes is a priority of researches. Their use in the building material and construction industry is a one of the possibilities which can help to keep the environment clean.

In the building material industry there are used a lot of types of wastes, which can be classified as follows:

- **By-product waste** is the waste produced by industry which includes any material that is rendered useless during a manufacturing process from plants, mills and mines. Usually they are storaged in landfills, which are placed on agricultural fields or around big cities. Some examples of industrial waste are silica fume, slag, sludge, fly ash, sand paper, metals, glass, etc. [1].

The by-products which are used in construction are:

- **Silica fume** is resulted from the processes of obtaining ferrosilicon industry, as a very fine powder which is recuperated by filters from furnaces. The quantity of dust involved by burnt gases from the furnace represents about 35% from the quantity of the end product. Silica fume generally contains more than 85% SiO_2, and also other components in smaller quantities, such as: Fe_2O_3 (1.3 - 4%), Al_2O_3 (0.85 - 2.5%), CaO (0.4 - 0.8%), Mg (0.6 - 1.5%), C

http://dx.doi.org/10.5772/59933

(1.1 - 2.5%). Silica fume has the shape of particles spherical and the specific surface is between 13000 and 23000 m^2/kg. Its spreading in the atmosphere has as effect the environment pollution [2].

- **Slag** is a co-product of the iron and steel production. Slag is usually a mixture of metal oxides and silicon dioxide. However, slags can contain metal sulfides and elemental metals. Slag is a valuable waste which can be used in agriculture, environment processes and construction industry [5]. In agricultural domain the slag is used for treatments for soil improvement. Other properties such as porosity, water holding capacity, bulk density make the slag suitable for using as adsorbent [4].

- **Sludge** refers to the residual material left from industrial wastewater or sewage treatment processes. It can also refer to the settled suspension obtained from conventional drinking water treatment and numerous other industrial processes [5]. This waste can be contaminated with toxic organic and inorganic compounds [4].

Another source of sludge and slag is from steel industry and they are generated as waste material or byproduct. They contain considerable quantities of valuable metals and materials. Different technologies are used for recovering the metallic parts, such as: classification, magnetic separation, leaching, roasting, etc. The wastes are then transformed in different sorts of waste, such as powder, conglomerate, etc. function the necessity of applied technology for a better use of natural resources and environment protection.

The paper industry that uses recycled paper as raw materials has as by-product paper sludge, which has a high content of calcium carbonate, organic materials and other minerals. Because its pozzolanic activity, the paper sludge can be used as cementitious materials in building industry.

- **Fly ash** is a residue from power plants or from different processes of incineration of solid materials. The fly ash is disposal on the landfill [6].

In Romania annually great fly ash (FA) quantities resulted: in 1980 resulted 15 million tons of FA, in 1985 obtained 30 million tons, and after 1990 the FA quantities decreased because the electricity consumes reduced. In our area annually resulted around 21740 tons of ashes (fly and bottom ashes). In the last twenty years resulted approximately 500 million tons of fly ash, from that a small part is capitalized. The FA unused is disposal on the landfill [7].

Toxic substances in the waste - including arsenic, mercury, chromium, and cadmium - can contaminate drinking water supplies and damage vital human organs and the nervous system. Ecosystems are also been damaged by the disposal of coal plant waste.

Fly ash produces environmental damage by causing air and water pollution on a large scale while the cost of storage of this waste is very high. The most serious problem is the hazard to atmosphere and underground water quality which would be a potential risk to the health and property of citizens and cause a huge stress to the economic and environmental system [7].

Another source of fly ash waste is from the solid waste incineration technology which is used in big cities of the world because its effectiveness in volume reduction, weight reduction and toxicity reduction, and also in energy and resource conservation. However, this technology produces fine fly ash residue which is equivalent to 2-5% of the original waste according to the Chinese researchers [8]. The municipal solid waste incineration (MSWI) fly ash can be used as raw material in sintering and preparing calcium sulphoaluminate cement (CSA), which had similar properties as the control cement [8].

- **Organic wastes** are generally biodegradable materials which are accumulated rapidly and for their storage it must design and realize great disposal landfills.

Biodegradable waste can be decomposed in a short period of time, under the natural conditions into the basic compounds, usually micro-organisms, bacteria, etc. This type of waste is found in municipal solid waste and is resulting from food, paper, biodegradable materials, etc. The wastes which are decomposed in the absence of oxygen are also considered as biodegradable waste and here are included wastes from manure, sewage, animal fat, palm fruit bunch, sugar bagasse, banana leaves, etc [1].

At the Iasi municipal sewage water treatment station, one of the greatest from Romania, a flow of 4.2 m^3/s has been processed daily, the sludge resulted by processing reaching an amount of about 3,600 t/day.

The fermentated sludge of the Iasi municipal treatment station had a neutral reaction and content in organic carbon of 29 - 34 %. The concentration of nitric nitrogen is low (0.16 - 0.42 ppm) and that of ammoniacal nitrogen between 24 and 830 ppm. The total content in macro-nutrients (N, P, K, Ca, Mg) from fermented sludge is 1.37 % N total, 1.19 % P total, 4.45 % K total, the calcium content is higher (3.12 %) and organic S and Mg have normal values comparable to those from soils.

The total content in heavy metals (Cd, Co, Cr, Cu, Fe, Mn, Ni, Pb, and Zn) of fermentated sludge is low, comparable to the one from soils, and under values of maximum admitted values in sludges, in order to be used as fertilizers for soils. Bacterial and fungic microflora of sludges has similar values to the ones of composts, size and composition of microorganism populations were very close to the ones of soils. The lack of bacteria from Salmonella group and a low number of coliform bacteria made a fertilizing material without contamination danger caused by pathogenic microflora from the fermentated sludge. Because the soils from the Moldavian area, most of them placed on slope lands, are poor in organic matter and nutrients, these sludge, adequately used, could be a substitute for a great part of expensive technological consumptions (mineral nutrients) and contribute to the improvement in organic matter content from soil.

- **Mineral wastes** are resulted from the industry processes where the natural resources are transformed in products. In construction industry a lot of natural raw materials are used in natural state. Mining, from the exploration to the closing stage, has a severe impact on the environment. Environmental impact can be direct through the activities: prospecting; exploration; site improvement; extraction; mineral preparations; mineral storage and preparation for delivery; transportation to beneficiary [9].

All technical stages in the minerals exploitation have an important impact on the environment and community life; the dust waste is very aggressive in the atmosphere, in water and soil because of the fine particle and toxic elements. The huge noise from technological processes is also an inconvenient for the community.

- **Inert waste** is waste which is neither chemically or biologically reactive and will not decompose in time. Examples of inert wastes are sand, drywall, and concrete. The inert waste typically requires lower disposal fees than biodegradable waste or hazardous waste.

In the industry of building materials the raw materials are used in natural state and as processed materials (case in which they are transformed in other materials (lime, cement, plaster, additives, etc). Among the raw materials there are: clay, calcareous, gypsum, dolomite, marble, mica, granite, etc. Natural aggregates are obtained from gravel from river or from quarry. Because the aggregates are used in different sorts as sizes, the natural aggregates are usually crushed for obtaining an imposed granulometry. A lot of powder waste remains after aggregate selection and their disposal affect the vegetation of the environment.

- **Agricultural wastes** are resulted from agricultural domain. They are biodegradable in time, but during the degradation process they must be storage in special places. For eliminating the wastes usually these wastes are burned, the powder resulted can be used as fine part in construction.

- **Construction demolition wastes** are resulted from new construction, rehabilitation or the demolition. As waste materials from construction can be: wood, drywall, masonry, metal, concrete, plastic, glass, cardboard. The construction waste quantities are bigger in metropolitan areas, where there are more buildings. Usually the construction waste is storage in landfills.

Construction wastes are obtained during the building process or after demolition. Different types of materials such as bricks, concrete, mortar, wood, steel rebar, insulation material, electrical wiring, plastic materials, glass, iron plate, tile, sanitary pieces, etc. which can be unused or damaged. According to specialty literature about 10% to 15% of materials are lost from the total building material, quantity which varies from site to site [4]. The uncontrolled disposal of this waste is very dangerous for the environment because building materials can contain toxic substances such as lead, asbestos, aluminum, etc.

The recycling demolished waste as aggregate in ordinary concrete offers a solution to the preservation of natural resources and the disposal of construction residues.

- **Transportation industry wastes** are represented by used tires, asphalt and concrete aggregate. Huge quantities of tires were used in artificial reefs, break waters, dock bumpers, soil erosion control mats, etc. [4].

3. Wastes from industry used in building materials

The concrete became a more interesting building material because it has improved its properties and also it is suitable for combining with different types of wastes. The presence of additions and/or fibers is also important because it can improve the performances of concrete or they allow the use of smaller quantities of cement.

The use of different waste in the concrete mix or for obtaining new types of concretes had as result the development of a new type of construction materials: green materials. In this category is included inorganic polymer concrete which is obtained predominantly from industrial waste materials. Concrete of any type had been used as it is or in combination with other materials, the most known being the steel with which had resulted reinforced concrete and prestressed concrete, that are still today very common and useful in construction industry. Polymer concrete is a new type of concrete in which the cement is replaced by a polymer. A high variety of waste are used for obtaining concretes of different requirements related to strength, to chemical resistance, with high durability, rapid hardening, etc.

An important way to use the wastes is to introduce them as a powder or filler in the composition of construction materials (cement, concrete, asphalt, etc.) or to use as aggregates (concrete or bricks from demolition can be used as an aggregate, steel slag can be transformed into aggregates, etc.). Concrete is one type of building material that can incorporate many types of waste such as silica fume, fly ash, cinder, husk, tires, glass, etc. Concrete is used for obtaining structural elements and constructions of any type.

3.1. By-product wastes

- **Silica fume** is specially used as mineral admixture in concrete because of the fineness of the particles which can fill better the spaces between the components of concrete mix. The new types of concretes (the high strength and high performance concrete, ultra-high performance concrete, with compressive strengths going to 150-180 MPa), high strength polymer concrete, etc. that are used in the new modern structures are obtained by adding in the mix silica fume in dosages between 8-12% [3]. Experimental studies shown that the compressive strength of concrete can increase with about 20% in the case of a dosage of 10% silica fume [10]. The behavior of different types of elements realized with high strength concrete under loading is improved, their resistant capacity is higher and the sizes of structural elements are reduced in comparison with structures realized of ordinary concrete.

In the ordinary cement concrete or polymer concrete, silica fume is added in different percentages, as replacement or not of cement, for improving the properties, in particular the compressive strength, durability characteristics, bond strength [10, 11]. These good effects of silica fume on the concrete are resulting from the fact that the particles of silica fume are very small and also from the pozzolana reaction of silica fume with cement paste components. In the behavior of structural elements it was observed according to experimental studies, that the failure of beams was improved, the concrete with silica fume had a better behavior to shear force, the number of cracks in tension zone at failure was reduced, which indicate that elements are less destroyed at failure [11].

In the hydraulic constructions, concrete with silica fume responds better to requirements of hydraulic construction because this concrete has a better behavior to frost-thaw cycles, to abrasion, cavitation, is resistant to chemical attack and it is less permeable, facts which result in a smaller dosage of cement.

Silica fume is also used for obtaining other types of concrete, such as self-compacting concrete, fiber reinforced concrete, polymer concrete. In the case of polymer concrete from experimental studies it was concluded that the increase in compressive strength is not too much as in the case of tensile strength.

Also the addition of silica fume decreased the content of polymer, which is an expensive material [2, 12]. The good behavior of concretes with silica fume can be used for realizing hybrid elements for constructions such as beam or columns, to which the tension zone is realized of polymer concrete, that has a better behavior in tension, and the compression zone of high strength concrete, having in view the better behavior of concrete in compression [11].

- **Slag** can be used in preparing composite cements or as aggregates in preparing concrete [12].

Slag cements are used in concrete structures because it gives some advantages, such as: less carbon dioxide emission, during the production, lower hydration heat during hardening, low permeability and good resistance to sulphate attack [13].

Ground granulated Blast Furnace Slag (GGBFS) improves the flexural strength and compressive strength of concrete and asphalt mixes, which recommend its use in roads, highways, pavements, hydraulic constructions, etc. Ground granulated slag is used in producing cement concrete as mix compound of the concrete or as component of cement. The use of ground granulated slag as component of concrete has the advantage of using it in different dosages, which is important in obtaining desired properties. Ground granulated slag can be used in obtaining Portland blastfurnace cement, which contains up to 5% until 95% of filler. Also, this type of waste can be used in preparing concrete as cementitious material due to its hydraulic property. In this case the fineness of ground granulated slag must appropriate to that of cement or even greater. The use of ground granulated slag used in obtaining concrete is benefic for the environment, but also it improves some properties of concrete such as: fresh concrete has a better workability, structure of hardened concrete is more compact, that resulting in increasing the long term strengths and durability. The content of ground granulated slag in the mix and its fineness depend on the purpose for which it is used in obtaining specific properties of concrete. Research studies reported a replacement of cement with dosages between 10 and 80% from the cement mass. The smaller quantities of waste are for increasing mechanical properties and high dosages are for improvement the resistance to chemical attack [14, 15].

The ground granulated blast furnace slag is also used in asphalt concrete for roads, highways, pavements, etc. An important utilization in the last time is to obtain high performance concretes, with improved durability, which is required in bridges, marine constructions, hydraulic dams, etc.

Another possibility of consuming ground granulated slag waste is to manufacture fibers which can be used in production of insulation material as slag wool [16].

Experimental studies on concrete with aggregates obtained of steel slag had shown that this type of waste can be used in road construction or in infrastructure works because the presence of steel increased the density of hardened concrete. Good mechanical properties were obtained in the case of cement concrete and polymer concrete with slag aggregates and addition of silica fume [3].

- **Sludge** is used in the production of concrete as filling material because its benefits such as improving the compressive strength, freeze-thaw resistance and waterproofness. Also it can be used as replacement of fine aggregates in asphalt paving [4].

The paper sludge is used for obtaining blended cements which contain 90% Portland cement and 10% waste. Also, the paper making waste can be processed to obtain a composition of cellulose fibers and clay which is suitable to use as insulating material or as filler in building materials [17, 18].

The utilization of paper waste sludge obtained from a paper industry, as a replacement to the mineral filler in various concrete mixes was experimentally analyzed [18]. Concrete mixes containing various contents of the waste (3, 5, 8 and 10%) were studied and the results shown a recommended replacement of sand of about 5% for obtaining concrete for masonry construction.

- **Fly ash**

The fly ash utilization is diversified in time and referring to construction industry this waste is used in: cement and concrete manufacturing, production of bricks, tiles and pavements, lightweight aggregates, etc. The new researches used fly ash in obtaining eco-concrete, which eliminated from the mix the cement, the geopolymer obtained being a material more friendlily with the environment. Although a large proportion of global FA is used by the building industry, there is a still proportion which is disposed of in ponds or landfills [4].

In the cement production the fly ash is used in the composition, in different quantities and the cement obtained are named composite cements [6].

In the cement-concrete production, a part of cement is replaced with different dosages of fly ash, normal dosages being between 10-40% and up to 75%. The advantages of using fly ash in concrete are given by the reduction of cement dosage, and also by the beneficial effects which improve concrete properties (mechanical strength and durability resistance), reduce bleeding, reduce cracking, decrease the heat during hardening of concrete [4, 19]. Experimental studies on cement concrete with fly ash shown that the addition of fiber, near fly ash is beneficial in improving the properties. Statistical optimization of mechanical properties for a concrete with 10% replacement of cement recommended for example for glass fiber type, a percentage of 1% from the concrete mass and a length of fiber of 35 mm in the case of compressive strength and higher percentages and smaller length, in the case of tensile strengths.

In obtaining the inorganic polymer concrete, which is a "green" material, fly ash that is considered alkali activated cement, replaces totally the cement from the mix. In fly ash-based geopolymer binder, fly ash reacts with an alkaline solution and the geopolymer paste acts as only binder for aggregates. The basic ingredients of fly ash-based geopolymer concrete are fly ash, sodium hydroxide, sodium silicate, fine aggregates and coarse aggregates [20].

The formulation of high-performance materials that are stronger and more durable than conventional cement-based materials has emerged as an issue of considerable importance in the construction industry. It is possible to utilize fly ash to produce a high-performance material at a potentially lower cost and without compromising its structural integrity. The high-performance polymer concrete made with fly-ash fillers presents the compressive strength, flexural strength, creep deformation and bond strength with values bigger than that of Portland cement concrete. Even in the case of fly ash the polymer dosage can be higher than in the case of other additions, the mechanical properties are increased in comparison with polymer concrete without addition. The use of fly ash as an aggregate in polymer concrete is very promising because it could be used as an overlay in pavement, bridges, and runways or in precast applications such as utility, transportation, and hydraulic components [21].

Industrial fly ash is also used for the production of low-strength material, also known as 'flowable fill'. It is used as a replacement of compacted soil in cases where the application of the latter is difficult or impossible. Also other wastes such as the cement kiln dust, asphalt dust, coal fly ash, coal bottom ash and quarry waste are used for preparing low-strength building materials. The content of these wastes in the mix is between 25-50%.

3.2. Organic wastes

One of the methods of consuming *sewage waste* was the obtaining of methanol gas or caloric fuel which is used for generation electrical powder. Another use is to transform it in powder and to use as fine addition in building materials.

The use of sewage sludge as an organic fertilizer has become of particular interest in the light of the EU Directive concerning the use of sewage sludge, which creates the need for cleaner production technology.

Sewage sludge can be an alternative for the protection of ecosystems. Firstly, sludge-borne organic matter is a crucial factor in improving aggregate stability and water holding capacity of soils, so that the risk of erosion may be reduced. Secondly, sludge-borne nutrients can make sewage sludge an excellent and cheap organic fertilizer for the crops [22]. However, the presence of inorganic and organic contaminants can hider such use of sewage sludge [23]. Moreover, it is well known that the application of organic materials to soil can sequester C, and thus contributes to the improvement of reduction of CO_2 in the atmosphere.

The sludge or the ashes obtained by burning the sludge can be used for obtaining ceramic products such as tile, brick block, pavement, etc. Several works have been carried out in this field [24]. The results from these works concluded that the use of treatment plant sludge as an additional component in a construction material, Portland cement concrete, is possible. The characteristics of sludge were evaluated. Also, it was necessary to analyze other properties such as the origin of the sludge, the components used, and the compatibility of the sludge within the cement matrix and the production of samples. Studies were conducted on the effect of sewage sludge ash on the workability of cement mortars.

A nonlinear reduction of workability in mortars containing sewage ash was observed. In their researches Monzo et. al. [24] reported the influence of sewage sludge ash (SSA) on the properties of cement mortar: a reduction of workability when a part of cement is replaced by sludge ash because it's higher water absorbtion characteristic. Studies on pozzolanic activity of SSA have shown that it contributed to the improved of compressed strength, but its effect is influenced by the sulfur content. The high sulfur content of sewage waste seems to have little influence on compressive strength of mortars containing sewage ash.

Moreover, Casanova et al. [25] observed that cement degradation processes had been observed when gypsum contaminated aggregates or sulfide-bearing aggregates are used in concrete mix.

The sludge or the ashes obtained by burning the sludge can be used for obtaining ceramic products such as tile, brick block, pavement, etc.

The sewage sludge ash can also is used as replacement of sand addition to brick clays which presented a high resistance to fire than normal brick clays.

Sewage sludge can be converted into slag, and as glass materials it is used to produce crystallized glass for ceramics technology.

From the environment point of view, the researchers Cenni at al. [26] studied the possibility of using fly ash resulted from co-firing of coal and sewage slag as additive in bilding materials, because the European standards forbid their use. Their studies shown that the ash from co-firing contained components such as unburned carbon, alkali, magnesium oxide, etc. with a reduced concentration as standard requirements. The authors required modifications in European standards for limiting elements that can be unfavorable by using them in building materials. Fytily and Zabaniotou [27] re-analyzed in a review article the use of sewage sludge in construction industry. Other use, such as incineration of sewage sludge is another way for consuming this waste, but it needs a rigorous control of gas compounds which depends on the technology that is used.

3.3. Mineral waste

The inert mineral waste resulted from quarries, from industrial processes can be used as aggregate or fine part in obtaining building materials and construction products. In any type of concrete these waste can replace different sorts of aggregates, contributing to preservation of natural resources.

The research studies in this domain shown that in building material industry a lot of inert waste can be used, such as granite, marble, limestone in the production of different materials: concrete, bricks, prefabricated elements, etc. The use of marble and granite waste in concrete preparing has shown that they improved the mechanical properties, workability and chemical resistance of concrete [28]. The polymer concrete with marble waste is of great interest because the marble addition or the marble used as aggregates improve the properties of concrete and contribute to a reduction of polymer content. The marble waste can also be used in the production of other building materials, such as ceramic products, where can be used as mix component, or in asphalt production as aggregate sort.

3.4. Construction demolition waste

The construction sector produces high quantities of wastes, over80% being solid waste which is dumped. Some of these wastes may have particular health, safety and environmental concern, such as, asbestos materials with lead-based paint coating and lighting waste. These materials are not included in the present review.

Until now, the construction practice was thought unsustainable because, not only it is consuming enormous quantities of stone, sand and drinking water, but also huge quantities of cement [4]. Modern reinforced concrete structures begin to deteriorate in 10 to 20 years. So, an important problem of concrete structures is that of increasing their durability. New types of concretes obtained by using Portland cement replacement materials and recycling the concrete removed from structures will contribute to the sustainability of building material industry [29]. Also, it must realize that the resources for construction industry are limited and the new technologies of obtaining building material must be based on the existing wastes.

The construction wastes are easier for recycling because they were parts of constructed buildings and as raw materials they were analyzed as raw materials. The concrete from demolition can be used as aggregate. Recycled-aggregate concrete is prepared by completely substituting of natural aggregates [30]. In many cases in the concrete mix there are also used superplasticizers and supplementary cementing materials (for example fly ash). Also, in the recycled-aggregate concrete mix, the cement can be replaces by fly ash or other by-product.

The other materials resulted from constructions such as wood, masonry, metal, plastic, fiber glass, polystyrene granules, etc. can be used in building industry. In the category of "green concrete" which means a concrete with waste, they are introduce in the mix different additions, some as filler. Cement concrete with wood waste is a concrete of low strength, and with characteristics of a lightweight concrete. In the case of cement concrete with polystyrene granules, experimental studies shown that the mechanical characteristics can be comparable with that of an ordinary concrete, even the density indicates a lightweight concrete [3]. A specific property of cement concrete with polystyrene is that of elastic behavior of material under loads, in the case of high dosages of polystyrene.

The concrete with polystyrene spheres was studied from a long time, and near the fact that it is a lightweight concrete other advantages recommend its use in construction. Concrete with polystyrene can be prepared in site or to obtain prefabricated units in factories. The properties of concrete with polystyrene are influenced by polystyrene dosage and by the size of granule. It has been shown that these properties can be significantly improved by adding steel fibers or additions (silica fume, fly ash, etc.) in the concrete matrix or by decreasing expanded polystyrene sphere size.

The polystyrene waste can be also used in manufacturing lightweight concrete blocks or surface units, with improved thermal insulating properties, by introducing the polystyrene sphere as lightweight aggregate in the concrete mix [33].

3.5. Transportation industry waste

The used tires are occupying a large landfill space and generate important problems to the society: one is that of hazard fire which is almost impossible to extinguish and the other is related to the people health. The European Association of Tyres and Rubber producers had estimated in 2009 that a quantity of 3.2 millions of tons of used tyres, from which 96% were re-used: 18% were retreated, 38% were recycled and 40% were used for burning in production of energy [32]. In Romania the recovery ratio is under 10%, in this context capitalization is a challenge for researchers.

The wastes of tyres are used in different purposes: for fixing and sealing soils in agricultural domain, in hydraulic domain (retaining walls, breakwaters), etc. [32].

The tyre waste can be used in natural form, cut in aggregates or in powder. Rubber aggregate is often used in construction works for obtaining light concrete or for road pavements. The experimental studies showed a percentage of around 25% from the mass for obtaining properties comparable with that of ordinary concrete. Higher quantities of tire waste result in decrease of mechanical properties [34]. The rubber increases the capacity to absorb energy from impacts, thus reducing the damage from collisions and increases the deformability and ductility of concrete. Rubber granulate is used for kindergarten play areas.

Once asphalt-rubber mixtures started to be regulated in the 1990s their use in pavements for roads and highways increased significantly [4]. The main advantages of pavements containing tires are their greater resistance to temperature variations and frost-thaw cycles, reduced noise, lower maintenance costs, a better drainage and an increase service life.

In different types of concrete the tire waste is used in various ways. In the concrete mix, the aggregates can be replaced by rubber particles in dosages between 0 - 45 % by volume. As indicated in the literature the concrete with used tire presented an decrease in mechanical characteristics, the use domain of materials obtained with this type of waste presents some advantages which derive from good damping properties, good thermal and acoustic performances [17].

In this direction, obtaining tough materials can be realized by introducing rubber particles in any mix. Concrete of any type is a brittle material. Small quantities of rubber in combination with other additions, can contribute to a better behavior of concrete, without affecting its mechanical properties.

Waste tire can be used as powder in obtaining cement concrete, polymer concrete, concrete with fibers, etc. Tire powder can be introduced in the mix as filler or to replace a part of fine aggregates. In the case of epoxy polymer concrete with powder of tire waste the experimental tests shown that the concrete is lightweight concrete with low mechanical properties, that recommend this concrete for pavement, prefabricated elements for sound protection, thermal insulation, etc [34]. Also this type of concrete showed a very good behavior to attack of chemical agents, abrasion resistance, so its use as floor in chemical industry or as pavements can be a possibility.

Waste tire can be used also in combination with other materials, such as glass fiber reinforced composite, in this case the tire waste being used as replacement of sand and for a better protection against pollution caused by noise. This composite can be used as façade panels for the cover of different buildings.

Rubber tires can be used in embakement as a lightweight filling material for soil reinforcement.

3.6. Other types of wastes used in building materials

- **The plastic materials** represent today an important category of waste. Most of them are re-used in different domains. Polypropylene, polyethylene, polyvinyl alcohol, polyvinyl chloride, nylon, aramid, polyesters are used as short plastic fibers in concrete elements. In the concrete production these fibers are currently used for obtaining high strength concretes, shotcrete, self-compacting concrete, etc [35]. Polyethylene terephthalate (PET) is one of the most used plastic in the entire world, especially for obtaining beverage containers, which are generally thrown away after single usage and their disposal creates serious problems to the environment. Some PET wastes are recycled for obtaining new products, other wastes are used as short fiber reinforcement in structural concrete, also as synthetic coarse aggregates for lightweight concrete, or as resin for polymer concrete [35].
- **The inorganic solid waste** can also be vitrified in solid-like glass materials that are used to manufacture aggregates for the construction industry for obtaining tile and bricks. No ashes are produced because at more than 5,000ºC, all the molecules are disintegrated.
- **Wastes of fiber** of different types (glass, polypropylene fiber, carbon, polyester, textile, etc.) and length are used din obtaining concrete with disperse reinforcement. The properties of fiber reinforced concrete depend on the fiber type, the geometry, the percentage of fiber, orientation and distribution of fiber, mixing and compaction of concrete.

The various applications of fiber reinforced concrete such as shotcrete in underground works, precast products, architectural panels, hydraulic constructions, etc. had contributed to the rapid development of this new building material.

3.7. Agricultural waste used in building material

Wastes from agricultural activities are in very high quantities, especially in some places of the world and they are another source of environment pollution and social problems because their accumulation in landfills and uncontrolled burning.

- **Rice husk** is generated by the rice milling process, from which 78% of weight is rice, broken rice and bran and the rest of 22% is husk. Some quantities of rice husk is burnt, which is polluting the environment. In the composition of rice husk there is nearly 20% silica, which after thermal treatment converts to a crystalline form that is with high reactivity, ultrafine size and large surface area. Because it's high pozzolanic activity the rice husk silica is used in obtaining high strength concrete instead silica fume. The cementing properties of rice husk offer the possibility of its use in ordinary concrete as cementitious material, for replacing cement or in production of supplementary cementing material [31]. Other uses are referred to its use as filler in polymer concrete, green concrete or in production of green building materials.

- **Banana leaves ashes** had been studied because it's pozzolanic activity which arises from the content of amorphous silica. The banana leaf ash is obtained by burning at a controlled temperature. The use as addition in mortar and concrete for civil structures had some advantages such as a reduction of costs of building materials and the consumption of huge quantities of banana produced every year [36].
- **Bamboo leaf** waste was experimentally analyzed because it's pozzolanic property which can be used for introducing this waste in cement composition. The test results shown that in composition of bamboo leaf waste the SiO_2 are 78.7%., being a very reactive pozzolan, comparable to silica fume. The blended cements obtained with bamboo leaf waste in a percentage of 10 and 20% showed the same compressive strength as the witness cement [37, 38].
- **Bagasse ash** is a waste sugar factory and it is used in obtaining blended Portland cements [39] or as replacement of cement in concrete in dosages of 10 to 30% of binder.

Natural cellulosic fibers can be used in the design and manufacturing of composite materials. The natural cellulosic fibers are bagasse from sugar cane, banana trunk from the banana plant and coconut coir from the coconut husk. The banana fiber exhibited the highest ash, carbon and cellulose content, hardness and tensile strength, while coconut the highest lignin content [40]. In combination with other additions, the concrete prepared with natural fibers exhibits good mechanical properties.

4. Conclusions

The useful life of a material in place, however, is always related to the particular combination of environmental factors to which it is subjected, so that durability, or service life, must always be related to the particular conditions involved.

In the last years, the degradation of environment is more pronounced because the measures of protecting the surrounding natural places were not respected and maybe, not understood. The increase of population, the huge energy which is needed, the development of industrial processes, all these resulted in a higher consumption of natural resources, more wastes and a higher pollution [41].

New building materials have developed because the new tendencies of obtaining eco-materials and protection of natural resources [42]. There are many years since when the cement industry has incorporated significant quantities of wastes (silica fume, fly ash, blast furnace slag, metakaolin, ceramic waste, etc.) because energetic, economic and environmental protection reasons. In recent years, alternative additions - bagasse ash, bamboo leaf ash, paper sludge, have been studied as components of eco-cements. The new generation of building materials which are combined with different types of wastes can offer a possibility of consuming disposal materials and reduce the environment pollution.

Also, the development of composite construction materials with low thermal conductivity using wastes will be an interesting alternative that would solve simultaneously energy and environment concerns [43].

The concrete of any type can be obtained by adding wastes, with experimental studies and statistical optimization, which help to characterize the new materials.

The developments in building materials must be sustainable and in the same time they ensure a ratio cost-energy that satisfy the modern requirements. The addition of wastes to concrete can improve or diminish some properties of the material. Therefore, a combination of wastes is often used or other materials are introduced into the composition to compensate for any disadvantages. These materials can be fibers of different types or lengths (steel, glass, polyester, carbon, bore, textile, etc.) or nanomaterials (nanotubes of carbon, nanoargillaceous materials, etc.).

New building materials based on nano-materials will develop and will influence the construction sector. Waste can be used for producing nanopowder or other nano-products which by using new nanotechnologies allow obtaining a new generation of cement based materials, more durable, with higher mechanical properties or even with desired properties, such as electrical conductivity as well as temperature, etc [44]. Today, nanotechnologies are in pre-exploration stage and must find application from experimental research to applications.

Construction composite materials are developing on the base of new researches in the recycling domain as an innovative option with environmental, economic and performance benefits.

Author details

Marinela Barbuta[1], Roxana Dana Bucur[2], Sorin Mihai Cimpeanu[3], Gigel Paraschiv[4] and Daniel Bucur[2]

1 "Gheorghe Asachi" Technical University in Iasi, Faculty of Civil Engineering and Building Services, Romania

2 University of Agricultural Sciences and Veterinary Medicine in Iasi, Romania

3 University of Agronomical Sciences and Veterinary Medicine in Bucharest, Romania

4 University Politehnica of Bucharest, Romania

References

[1] Maczulak Anne Elizabeth. Pollution: Treating Environmental Toxins. New York: InfoBase Publishing. p. 120; 2010.

[2] Abul Bari, H., Tabet Abid, R., Mohammad, A.H. Fume Silica Base Grease. Journal of Applied Science 2008; 8(4), 687-691.

[3] Barbuta Marinela. Effect of different types of superplasticizer on the properties of high strength concrete incorporating large amounts of silica fume, Bulletin of the Polytechnic Institute of Iasi 2005; 51(1-2), 69-74.

[4] Bolden J., Abu-Lebdeh T., Fini E. Utilization of recycled and waste materials in various construction applications. American Journal of Environmental Science 2013; 9(1) 14-24 doi: 10.3844/ajessp.2013.14.24.

[5] McBride, M.B. Toxic metals in sewage sludge-amended soils: Has promotion of beneficial use discounted the risks? Advances in Environmental Research 2003; 8 5-19. doi:10.1016/S1093-0191(02)00141-7

[6] Dai, S., Zhao L., Peng S., Chou C.L., Wang X., Zhang Y., Li D., Sun Y. Abundances and distribution of minerals and elements in high-alumina coal fly ash from the Jungar Power Plant, Inner Mongolia, China. International Journal of Coal Geology 2010; 81, 320-332.

[7] Harja Maria, Barbuta Marinela, Gavrilescu Maria. Study of morphology for geopolymer materials obtained from fly ash. Environment Engineering Management Journal 2009b; 8, 1021-1027.

[8] Xiaolu Guo, Huisheng Shi, Wenpei Hu, Kai Wu. Durability and microstructure of CSA cement-based materials from MSWI fly ash. Cement & Concrete Composites 2014; 46 26-31.

[9] Twerefou D.K. Mineral Exploitation, Environmental Sustainability and Sustainable Development in EAC, SADC and ECOWAS Regions, ATPC, Work in Progress 2009; 79, 1-12.

[10] Barbuta Marinela, Nour Doina Smaranda. Components compatibility to high strength concrete with silica fume, International Conference VSU Sofia. Tom I P.II 46-52; 2006.

[11] Barbuta Marinela, Mihai P. Behavior of reinforced concrete beam with concrete containing different amount of silica fume. International Conference Life Cycle Assessment, behavior and properties of concrete, Brno 16-22; 2004.

[12] Gao J. M., Qian C.X, Wang B., Morino K. Experimental study on properties of polymer modified cement mortars with silica fume, Cement and Concrete Research 2002; 32(1) 41-45.

[13] Bilim C., Atis C.D., Tanyildizi H., Karahan O. Predicting the compressive strength of ground granulated blast furnace slag concrete using artificial neural network, Advanced in Engineering Software 2009; 40, 334-340 doi: 10.1016/j.advengsoft.2008.05.005.

[14] Wei Y., Hansen W. Early-age strain-stress relationship and cracking behavior of slag cement mixtures subject to constant uniaxial restraint. Construction and Building Materials 49(2013), 635-642. doi: 10.1016/j.conbuildmat.2013.08.061

[15] Fredericci C., Zanotto E.D., Ziemath E.C. Crystallization mechanism and properties of a blast furnace slag glass, Journal of Non-Crystalline Solids 2000; 273 (1-3): 64-75. doi:10.1016/S0022-3093(00)00145-9.

[16] Mangat P. S., Khatib J. M. Influence of Fly Ash, Silica Fume, and Slag on Sulfate Resistance of Concrete. Materials Journal 1995; 92 (5) 542-552 doi: 10.14359/9775.

[17] Moriconi G. Recyclable in concrete technology: sustainability and durability. In: Rudolph N., Kraus, Tarun R. Naik, Peter Claisse, Sadeghi-Pouya, ed. Proceedings of International Conference: Sustainable construction materials and technologies, 11-13 June 2007 Coventry, Special papers proceedings. Publisher UW Milwaukee CBU, 1-12; 2007.

[18] Anderson M., Skerratt R. G., Thomas J. P., Clay S. D. Case study involving using fluidised bed incinerator sludge ash as a partial clay substitute in brick manufacture, Water Science and Technology, 1996; 34(3-4), (1996) 507-515 doi: 10.1016/0273-1223(96)00618-X.

[19] Magureanu C., Negrutiu C. Performance of concrete containing high volume coal fly ash - green concrete. International Conference *on* Computational methods *and* Experiments *in* Materials Characteristics. New Forest, Ukraine, June 17 - 19, Proceeding Papers. 64: 373-79; 2009.

[20] Sofi M., van Deventer J.S.J., Mendis P. A., Luckey G.C. Engineering properties of inorganic concretes (IPCs), Cement and Concrete Research 2007; 37, 251-257 doi: 10.1016/j.cemconres.2006.10.008.

[21] Gencel O, Koksal F., Ozel C., Brostow W. Combined effect of fly ash and waste ferrochromium on properties of concrete, Construction and Building Materials 2012; 290 633-640.

[22] Ailincai C., Jitareanu G., Bucur D., Ailincai Despina. Influence of sewage sludge on maize yield and quality and soil chemical characteristics, Journal of Food, Agriculture & Environment 2007; 5 (1) 310-313.

[23] Andersen A. Growth control of organism, i.e. yeast for brewing beer, in nutrient medium, involves monitoring added or produced gaseous materials in exhaust gas of medium, Patent Number(s): WO200155295-A (2001)

[24] Monzo J., Paya J., Borrachero M.V., Girbes I. Reuse of sewage sludge ashes (SSA) in cement mixtures: the effect os SSA on the workability of cement mortars. Waste Management 2003; 23 373-381 doi:10.1016/S0956/053X/(03)00034-5.

[25] Casanova I., Aggulo L., Aguado A. Aggregate expansivity due to sulphide oxidation-1. Reaction system and rate model. Cement Concrete Research 1996; 26(7) 993-8.

[26] Cenni R., Janisch B., Spliethoff H., Hein K. Legislative and environmental issues on the use of ash from coal and municipal sewage sludge co-firing as construction material. Waste Management 2001; 21 17-31.

[27] Fytili D., Zabaniotou A. Utilization of sewage sludge in EU application of old and new methods - A review. Renewable and Sustainable Energy Reviews 2008; 12, 116-140 doi:10.1016/j.rser.2006.05.014

[28] Hebhoub H., Aoun H., Belachia M., Houari H., Ghorbel E. Use of waste of marble aggregates in concrete, Construction and Building Materials 2011; 25 1167-1171 doi: 10.1016/j.conbuildmat.2010.09.037.

[29] Rao A., Jha K. N., Misra S. Use of aggregates from recycled construction and demolition waste in concrete. Conservation and Recycling 2007; 50(1) 71-81 doi: 10.1016/ j.resconrec.2006.05.010.

[30] Topcu I. B., Sengel S. Properties of concrete produced with waste concrete aggregate. Cement and Concrete Research 2004; 34(8) 1307-1312 doi: 10.1016/ j.cemconres. 2003.12.019.

[31] Khan R., Jabbar A., Ahmad I., Khan W., Khan A.N., Mirza J. Reduction in environmental problems using rice-husk ash in concrete. Construction and Building Materials 2012; 30 360-365 doi: 10.1016/j.conbuildmat.2011.11.028

[32] Bravo M., de Brito J. Concrete made with used tyre aggregate: durability - related performance. Journal of Cleaner Production 2012; 25 42-50 doi: 10.1016/ j.jclepro. 2011.11.066.

[33] Al-Jabri, K.S. Concrete block for thermal insulation in hot climate, Cement and Concrete Research, 2005, 35 (8), 1472-1479

[34] Diaconescu Rodica Mariana, Barbuta Marinela, Harja Maria. Prediction of mechanical properties of polymer concrete with tyre rubber using neural network. Materials Science and Engineering B - Advanced Functional Solid-State Materials 2013; 178 (19) 1259-1267 doi: http://dx.doi.org/10.1590/1516-1439.210413.

[35] Kim S.B., Yi N.H., Kim H.Y., Kim J.H.J., Song Y.C. Material and structural performance evaluation of recycled PET fiber reinforced concrete, Cement & Concrete Composites 2010; 32 232-240

[36] Kanning R.C., Portella K.F., Braganca M.M., Bonato M.M., dos Santos-Banana Jeannette. Banana leaves ashes as pozzolan for concrete and mortar of Portland cement, Construction and Building Materials 20041; 54 460-465 doi: 10.1016/j.conbuildmat. 2013.12.030

[37] Frias M., Savastano H., Villar E., Sanchez de Rojas M. I., Santos S. Characterization and properties of blended cement matrices containing activated bamboo leaf wastes. Construction and Building Materials 2012; 34 1019-1023 http://dx.doi.org/10.1016/ j.cemconcomp.2012.05.005.

[38] Singh N.B., Das S.S., Singh N.P., Dwivedi V.N. Hydration of bamboo leaf ash blended Portland cement. Indian Journal of Engineering & Materials Sciences 2007; 14, 69-76.

[39] Singh N.B., Singh V.D., Rai S. Hydration of bagasse ash-blended Portland cement. Cement and Concrete Research 2000; 30 1485-8.

[40] Justiz-Smith N., Virgo G.J., Buchanan V. E. Potential of Jamaican banana, coconut coir and bagasse fibres as composite materials, Materials Characterization 2008; 59(9) 1273-1278.

[41] Altan Dombaycı O. The environmental impact of optimum insulation thickness for external walls of buildings, Building and Environment 2007; 42 3855-3859 doi: 10.1016/j.buildenv.2006.10.054.

[42] Barbuta Marinela, Toma I.O., Harja Maria, Toma A.M., Gavriloaia C. Behavior of short hybrid concrete columns under eccentric compression. Archives of Civil and Mechanical Engineering 2013; 13 119-127.

[43] Benazzouk A., Douzane O., Mezreb K., Laidoud B., Que'Neudec M. Thermal conductivity of cement composites containing rubber waste particles: Experimental study and modelling. Construction and Building Materials 2008; 22 573-579. http://dx.doi.org/10.1016/j.conbuildmat.2006.11.011

[44] Sobolev K., Gutiérrez M.F. How nanotechnology can change the concrete world: Part II. American Ceramic Society 2005; 84(11) 16-19.

Chapter 4

Soil Salinity Control in Irrigated Land with Agricultural Drainage Systems

Carlos Chávez, Carlos Fuentes,
Fernando Brambila and Nami Morales-Durán

Additional information is available at the end of the chapter

http://dx.doi.org/10.5772/59370

1. Introduction

Soil salinity is a worldwide problem as well as in Central and Northern Mexico. Nearly 8.4 million ha worldwide are affected by soil salinity and alkalinity, of which about 5.5 million ha are waterlogged [1]. The problem worsens in arid and semiarid areas, in soils with insufficient drainage [2] and high evaporation [3]. In Mexico, there are 6.46 million ha irrigated mainly in the center and north territory areas [4]; 10-30 % of irrigated land is affected by salinity and nearly two thirds of this area is located in the North [5].

The salinization of these irrigated areas is an increasing problem and the lands are abandoned; therefore, a technical and economic alternative to recover this land is needed. Agricultural subsurface drainage is a solution which takes into account the technology by environment interaction, as well as lowering the water table levels along with the salt concentration in the soil profile [1].

The dynamics of water drainage systems has been studied by applying the Boussinesq equation [6] for unconfined aquifers using the finite element technique [7,8] and the finite difference [9,10], and the solutes dynamics has been studied by applying the Fick's law [11,12,13]. These results in the advection-dispersion equation, namely by gravity and Fick's law.

The solutes are also found in the gas phase and adsorbed by soil in the solid phase, the first phase is disregarded for purposes of transport modeling in water, but it is really important in terms of the amount of fertilizer transferred into the atmosphere at a given time [14,15,16], and incorporating the adsorbed substance in the solid phase. The relationship between the substance which transported by the water flow and the substance which adsorbed and exchanges in the soil solid structure is known as the adsorption isotherm [11,12,13].

A large number of models for simulating solute transport in the unsaturated zone are now increasingly being used for a wide range of applications in both research and management [17], some of the more popular models include SWAP [18], HYDRUS-1D [19], STANMOD (STudio of ANalytical MODels) [20], UNSATH [21] and COUP [22] but the majority of applications for water flow in the vadose zone requires a numerical solution of the Richards equation [23], also requires more calculation time in order to find the equation solution.

This study aims to solve the one-dimensional advection-dispersion equation using the technique of finite differences, coupled with the Boussinesq equation in order to model the transport of solutes in subsurface drainage systems, assuming that the solute is concentrated in the liquid phase.

2. Theory

2.1. The Boussinesq equation

In the study of the water dynamics in agricultural subsurface drainage systems using the Boussinesq equation, the variations in hydraulic head along the drain pipes (direction y) are negligible with respect to head variations in the cross section (direction x). It is the one-dimensional Boussinesq equation which is a result of the continuity equation, $\partial(\upsilon H)/\partial t+\partial(Hq)/\partial x=R_w$, and the Darcy's law, $q=-K_s \partial H/\partial x$, namely:

$$\mu(H)\frac{\partial H}{\partial t}=\frac{\partial}{\partial x}\left[T(H)\frac{\partial H}{\partial x}\right]+R_w \tag{1}$$

where $\mu(H)$ is the storage capacity, $H=H(x,\ t)$ is the elevations of the free surface or hydraulic head above the impervious layer $[L\ ]$, and is a function the horizontal coordinate (x) and the time (t), $T(H)$ is the transmissivity given by $T(H)=K_s H\ [L\ ^2 T^{-1}]$, R_w is the volume of recharge in the unit of time per unit of the aquifer $[L\ ^3]$, $\upsilon=\upsilon(H)$ is the drainable porosity as a head function, and K_s is the saturated hydraulic conductivity $[L\ T^{-1}]$.

The storage capacity, see [24], is: $\mu(H)=\theta_s-\theta(H-H_s)$, where θ_s is the saturated volumetric water content $[L\ ^3 L\ ^{-3}]$, and $\theta(H-H_s)$ represents the water content evolution in the position $z=H_s$, while the free surface decreases, and z is the elevation of ground surface $[L\ ]$.

2.2. The drainable porosity

To calculate the storage capacity and the drainable porosity it is necessary to provide the soil water retention curve. The model of van Genuchten [25] was accepted in field and laboratory studies: $\theta(\psi)=\theta_r+(\theta_s-\theta_r)[1+(\psi/\psi_d)^n]^{-m}$, where ψ is the soil water potential defined by $\psi=(H-z)$ $[L]$, ψ_d is the pressure scale parameter $[L]$, θ_s is the saturated volumetric water content $[L^3 L^{-3}]$, θ_r is the residual volumetric content $[L^3 L^{-3}]$, m and n are parameters (dimensionless) that determine the shape of the soil water retention curve. The introduction of this equation in the storage capacity results in the following expression for storage capacity: $\mu(H)=(\theta_s-\theta_r)\{1-[1+((H_s-H)/|\psi_d|)^n]^{-m}\}$.

The saturated volumetric water content can be assimilated to the soil porosity (ϕ), dimensionless, this is calculated with the formula $\phi=1-\rho_t/\rho_o$, where ρ_t is the bulk density $[M L^{-3}]$ and ρ_o is the particles density $[M L^{-3}]$; the residual volumetric water content (θ_r) is considered to be zero.

2.3. Initial and boundary conditions

To study the agricultural drainage with equation (1), the initial and boundary conditions should be defined at the domain. The initial condition is established from the water table position at the initial time. Dirichlet and Neumann boundary type conditions can be used on drains to solve equation (1), the pressure head on the drains is required in the first condition whereas the drainage flux is required in the second one [8]. A third type of boundary condition is a linear combination of the precedent conditions; this condition includes a resistance parameter to the flow at the soil–drain interface. Null resistance corresponds to the Dirichlet condition and infinite resistance corresponds to Neumann condition. The third condition is a radiation type condition [26]. In the case of drainage, the radiation condition establishes that drainage flux is directly proportional to the pressure head on the drain and inversely proportional to the resistance in the interface between soil and the drainpipe wall in concordance to the Ohm law.

The hydraulic head measured above the impermeable barrier $H(x, t)$ is associated with the head $h(x, t)$ measured from above the drains using: $H(x, t)=D_o+h(x, t)$, where D_o is the distance from the impermeable barrier to the drains $[L]$. Transversal variation of h at the beginning is considered as the initial condition $h(x, 0)=h_s(x)$, where h_s is the head on the drain in the initial time $[L]$. The fractal radiation condition for the Boussinesq equations is given by [8]:

$$-K_s\frac{\partial h}{\partial x}\pm q_s\left(\frac{h}{h_s}\right)^{2s}=0;\ x=0,Lx=0,L \tag{2}$$

where the positive sign corresponds to $x=0$ and the negative sign to $x=L$. L is the distance between drains; q_s is the corresponding flux to h_s and it is function of the soil-drain interface characteristic $[L\,T^{-1}]$. For the s parameter, the authors argued that it is defined by $s=D/E$, where D is the effective fractal dimension to the soil-drain interface, and $E=3$ is the Euclidean dimension of physical space. The relation of the s parameter and effective porosity is obtained from the equation $(1-\phi)^s + \phi^{2s} = 1$ given by [27]. Equation (2) contains as particular cases the lineal radiation condition when $s=1/2$ and the quadratic radiation condition when $s=1$. In a system of parallel drains, the drained water flows by length unit at each drain is: $Q_d(t) = 2[D_o + h(0,t)]q_s[h(0,t)/h_s]^{2s}$, and the cumulative drained depth is calculated by $\ell(t) = \frac{1}{L}\int_0^t Q_d(\bar{t})d\bar{t}$, where $\bar{t}$ is the integration variable.

2.4. Solute transport equation

The advection-dispersion equation used to study the solute transport [28,29,30], in a one-dimensional form, is a result of the continuity equation, $\partial(HC_T)/\partial t + \partial Qs/\partial x = R_s$, and the dynamic law given by $Qs = HqC - \upsilon HDa(\partial C/\partial x)$, namely:

$$\frac{\partial(HC_T)}{\partial t} + \frac{\partial(HqC)}{\partial x} = \frac{\partial}{\partial x}\left[\upsilon HDa\frac{\partial C}{\partial x}\right] + R_s \tag{3}$$

where Da is the diffusion coefficient in the water $[L^2T^{-1}]$; C_T is the total solute concentration in soil $[ML^{-3}]$; C is the solute concentration in water $[ML^{-3}]$; and R_s is the term which includes gains or losses of the solute due to chemical reactions and the extraction plant $[M]$. Note that q and υ are obtained from the water flow model. The diffusion coefficient in the water is calculated by $Da = \lambda v$, where λ is the dispersivity $[L]$ and v the interstitial velocity of water calculated by $v = q/\upsilon$ $[L\ T^{-1}]$.

The water soluble compounds which have a negligible vapor pressure can exist in three phases of soil: 1) dissolved in water, 2) as vapor in the soil atmosphere and 3) as stationary phase adsorbed to soil organic matter or in the clay mineral surfaces [11,12,13]. The total concentration of the compound (C_T), expressed in units of mass per volume of soil can be written as: $C_T = \upsilon C + \rho_t C_a$, where C_a is the concentration of the adsorbed compound $[ML^{-3}]$ and is a function of the concentration of the solute in the mobile phase (C_d) $[ML^{-3}]$ and the adsorption constant of the solute to the stationary phase surface (κ), $C_a = \kappa C_d$, namely linear isotherm. Thus, the concentration of the substance compared to the volume of the porous medium (C_T) will be the result of a part that is in the water, air and the dynamic equilibrium with the phase that generates it. Generally, in studies in small time scales, such as irrigation and drainage in a porous medium, the gas phase is not considered [29]. Thus, in this work, the concentration in the adsorbed phase, the concentration in the gas phase and the term R_s are ignored.

2.5. Numerical scheme

The numerical scheme used is based on the assumption that the solute is concentrated mainly in the liquid phase. Thus, the advection-dispersion equation in one-dimensional is given by equation (3). To solve this equation, we use the same discretization scheme to transfer water in the Boussinesq equation [10], for which two interpolation parameters are introduced: $\gamma = (x_{i+\gamma} - x_i / x_{i+1} - x_i)$ and $\omega = (t_{j+\omega} - t_j / t_{j+1} - t_j)$, where $0 \le \gamma \le 1$ and $0 \le \omega \le 1$; $i = 1, 2, \ldots$ and $j = 1, 2, \ldots$ are the space and time indices, respectively.

The dependent variable (Φ) in an intermediate node $i + \gamma$ for all j is estimated as:

$$\Phi_{i+\gamma}^{j} = (1-\gamma)\Phi_i^j + \gamma\Phi_{i+1}^{j} \tag{4}$$

while the intermediate time $j + \omega$ for all i is estimated as:

$$\Phi_i^{j+\omega} = (1-\omega)\Phi_i^j + \omega\Phi_i^{j+1} \tag{5}$$

The discretization of the temporal derivative in the equation (3) is:

$$\left.\frac{\partial(\upsilon HC)}{\partial t}\right|_i^{j+\omega} = \frac{(\upsilon H)_i^{j+1} C_i^{j+1} - (\upsilon H)_i^j C_i^j + (\rho_t H)_i^{j+1} C_{di}^{j+1} - (\rho_t H)_i^j C_{di}^j}{\Delta t_j} = b_2 C_i^{j+1} - b_1 C_i^j + b_0;\ \Delta t_j = t_{j+1} - t_j \tag{6}$$

where:

$$b_0 = \frac{(\rho_t H)_i^{j+1} C_{di}^{j+1} - (\rho_t H)_i^j C_{di}^j}{\Delta t_j};\ b_1 = \frac{(\upsilon H)_i^j}{\Delta t_j};\ b_2 = \frac{(\upsilon H)_i^{j+1}}{\Delta t_j} \tag{7}$$

The spatial derivative discretization in the continuity equation is:

$$\left.\frac{\partial Qs}{\partial x}\right|_i^{j+\omega} = \frac{Qs_{i+\gamma}^{j+\omega} - Qs_{i-(1-\gamma)}^{j+\omega}}{\Delta x_i};\ \Delta x_i = (1-\gamma)(x_i - x_{i-1}) + \gamma(x_{i+1} - x_i) \tag{8}$$

According with the dynamic law:

$$Qs\big|_{i+\gamma}^{j+\omega} = (Hq)_{i+\gamma}^{j+\omega} C_{i+\gamma}^{j+\omega} - (\upsilon H)_{i+\gamma}^{j+\omega} (Da)_{i+\gamma}^{j+\omega} \frac{C_{i+1}^{j+\omega} - C_i^{j+\omega}}{x_{i+1} - x_i} \tag{9}$$

$$Qs\big|_{i-(1-\gamma)}^{j+\omega} = (Hq)_{i-(1-\gamma)}^{j+\omega} C_{i-(1-\gamma)}^{j+\omega} - (\upsilon H)_{i-(1-\gamma)}^{j+\omega} (Da)_{i-(1-\gamma)}^{j+\omega} \frac{C_i^{j+\omega} - C_{i-1}^{j+\omega}}{x_i - x_{i-1}} \tag{10}$$

According with the equation (4), the spatial interpolation is:

$$C_{i+\gamma}^{j}=\left(1-\gamma\right)C_{i}^{j}+\gamma C_{i+1}^{j};\; C_{i-(1-\gamma)}^{j}=\left(1-\gamma\right)C_{i-1}^{j}+\gamma C_{i}^{j} \tag{11}$$

and according with the equation (5) the temporal interpolation is $C_i^{j+\omega}=(1-\omega)C_i^j+\omega C_i^{j+1}$.

The dependent variables involved in the advective term of the equations (9) and (10) are defined by:

$$C_{i+\gamma}^{j+\omega}=\left(1-\omega\right)C_{i+\gamma}^{j}+\omega C_{i+\gamma}^{j+1}=\left(1-\omega\right)\left[\left(1-\gamma\right)C_{i}^{j}+\gamma C_{i+1}^{j}\right]+\omega\left[\left(1-\gamma\right)C_{i}^{j+1}+\gamma C_{i+1}^{j+1}\right] \tag{12}$$

$$C_{i-(1-\gamma)}^{j+\omega}=\left(1-\omega\right)C_{i-(1-\gamma)}^{j}+\omega C_{i-(1-\gamma)}^{j+1}=\left(1-\omega\right)\left[\left(1-\gamma\right)C_{i-1}^{j}+\gamma C_{i}^{j}\right]+\omega\left[\left(1-\gamma\right)C_{i-1}^{j+1}+\gamma C_{i}^{j+1}\right] \tag{13}$$

while the dependent variables involved in the dispersive term of the same equations are defined by:

$$C_{i+1}^{j+\omega}=\left(1-\omega\right)C_{i+1}^{j}+\omega C_{i+1}^{j+1};\; C_{i}^{j+\omega}=\left(1-\omega\right)C_{i}^{j}+\omega C_{i}^{j+1};\; C_{i-1}^{j+\omega}=\left(1-\omega\right)C_{i-1}^{j}+\omega C_{i-1}^{j+1} \tag{14}$$

Equation (8) considering equations (9) and (10) can be written as:

$$\left.\frac{\partial Qs}{\partial x}\right|_{i}^{j+\omega}=a_1 C_{i+\gamma}^{j+\omega}-a_2\left(C_{i+1}^{j+\omega}-C_{i}^{j+\omega}\right)-a_3 C_{i-(1-\gamma)}^{j+\omega}+a_4\left(C_{i}^{j+\omega}-C_{i-1}^{j+\omega}\right) \tag{15}$$

where:

$$a_1=\frac{\left(Hq\right)_{i+\gamma}^{j+\omega}}{\Delta x_i};\; a_2=\frac{\left(\upsilon H\right)_{i+\gamma}^{j+\omega}\left(Da\right)_{i+\gamma}^{j+\omega}}{\Delta x_i\left(x_{i+1}-x_i\right)};\; a_3=\frac{\left(Hq\right)_{i-(1-\gamma)}^{j+\omega}}{\Delta x_i};\; a_4=\frac{\left(\upsilon H\right)_{i-(1-\gamma)}^{j+\omega}\left(Da\right)_{i-(1-\gamma)}^{j+\omega}}{\Delta x_i\left(x_i-x_{i-1}\right)} \tag{16}$$

Substituting equations (12)-(14) in equation (15) and associating similar terms allows obtaining:

$$\begin{aligned}\left.\frac{\partial Qs}{\partial x}\right|_{i}^{j+\omega}&=-\omega\left[a_4+\left(1-\gamma\right)a_3\right]C_{i-1}^{j+1}+\omega\left[\left(1-\gamma\right)a_1+a_2-\gamma a_3+a_4\right]C_{i}^{j+1}+\omega\left[\gamma a_1-a_2\right]C_{i+1}^{j+1}\\&-\left(1-\omega\right)\left[a_4+\left(1-\gamma\right)a_3\right]C_{i-1}^{j}+\left(1-\omega\right)\left[a_4-\gamma a_3+a_2+\left(1-\gamma\right)a_1\right]C_{i}^{j}\\&+\left(1-\omega\right)\left[\gamma a_1-a_2\right]C_{i+1}^{j}\end{aligned} \tag{17}$$

Substituting equations (6) and (17) in the continuity equation, the following algebraic equations system is obtained:

$$As_i C_{i-1}^{j+1} + Bs_i C_i^{j+1} + Ds_i C_{i+1}^{j+1} = Es_i;\ i = 2,3,\ldots,n-1 \tag{18}$$

where

$$As_i = -\omega\left[a_4 + (1-\gamma)a_3\right] \tag{19}$$

$$Bs_i = \omega\left[(1-\gamma)a_1 + a_2 - \gamma a_3 + a_4\right] + b_2 \tag{20}$$

$$Ds_i = \omega\left[\gamma a_1 - a_2\right] \tag{21}$$

$$\begin{aligned} Es_i = Rs_i^{j+\omega} &+ (1-\omega)\left[a_4 + (1-\gamma)a_3\right]C_{i-1}^j \\ &-\left\{(1-\omega)\left[a_4 - \gamma a_3 + a_2 + (1-\gamma)a_1\right] - b_1\right\}C_i^j - (1-\omega)\left[\gamma a_1 - a_2\right]C_{i+1}^j - b_0 \end{aligned} \tag{22}$$

The water flow and the head are obtained from the Boussinesq equation solution, so that they should be included in the system (18). To find the solution of the water transfer equation, it is necessary to specify the initial and boundary conditions, equation (18) can be solved with the Thomas Algorithm, see [31,10].

The Thomas algorithm, also known as the tridiagonal matrix algorithm (TDMA), is a simplified form of Gaussian elimination that can be used to solve *tridiagonal* matrix systems (equation 18) [32]. It is based on LU decomposition in wich the matrix system $Mx = r$ where L is a lower triangular matrix and U is an upper triangular matrix. The system can be efficiently solved by setting $Ux = p$ and then solving first $Lp = r$ for p and then $Ux = p$ for x. The Thomas algorithm consist of two steps. In the first step decomposing the matrix into $M = LU$ and solving $Lp = r$ are accomplished in a single downwards sweep, taking us straight from $Mx = r$ to $Ux = p$. In the second step the equation $Ux = p$ is solved for x in an upwards sweep [33].

2.6. Linear radiation condition

The radiation boundary condition, or mixed condition, is used to accept a linear variation between the dispersive flux and concentration difference with the external medium (C_{ext}) and the border, for all time. The linear radiation condition is due originally to Newton, who postulated that the heat flow at the border of a body is proportional to the temperature difference between the body and the medium that surrounds it; the result is equivalent to a Ohm law into electricity. To linearize these conditions, we introduce a generalization of the dimensionless conductance coefficient (κ_s), as follows: $-(\partial C/\partial x) + \kappa_s(C - C_{ext}/L) = 0$. If we observe the one-dimensional equation of solute transport, the dimensionless conductance coefficient (κ_s) must be zero by the advective component, however, the solution is allowed only for purposes of illustration to derive the boundary conditions.

2.7. Selection of the space (Δx) and time (Δt) increments

In reference [10] the authors discuss the selection of spatial and temporal increments pointing out a comparison of the depletion of the free surface for all time between the results obtained with the finite difference solution of the Boussinesq equation and the results obtained with an analytical solution reported in the literature. The same authors [10] concluded that the optimal interpolation that minimizes the sum of the squares errors are $\gamma = 0.5\Delta x$ (cm) and $\omega = 0.98\Delta t$ (h), for space and time respectively.

3. Application

3.1. Laboratory experiment

To evaluate the descriptive capacity of the numerical solution, a drainage experiment was conducted in a laboratory. The drainage module (see Figure 1) is the one used by [8] and [10]. The module dimensions are: $L = 100$ cm, $H_s = 120$ cm and $D_0 = 25$ cm. The drain diameter is $d = 5$ cm and the drain length is $\ell = 30$ cm.

Figure 1. Drainage module

The module was filled with altered sample of salty soil of Celaya, Guanajuato, México (see Figure 2). Soil passed through a 2 mm sieve and was disposed on 5 cm thick layers, in order to maintain the bulk density at a constant value. The soil was saturated by applying a constant water head (no salt) on its surface until the entrapped air was virtually removed. Once the drains were closed, the water head was removed from the soil surface; the surface of the module was then covered with a plastic in order to avoid evaporation. Finally, the drains were opened to measure the drained water volume; the initial condition was equivalent to $h(x, 0) = h_s$ and the recharge was null $R_w = 0$ during the drainage phase. Soil porosity (ϕ) was calculated with the formula $\phi = 1 - \rho_t / \rho_o$ (the bulk density was determined by the weight and volume of the soil of drainage module $\rho_t = 1.14 \, g / cm^3$ and the particles density $\rho_s = 2.65 \, g / cm^3$, $\phi = 0.5695$ cm^3 / cm^3 was obtained). The soil fractal dimension obtained was equal to 0.7026.

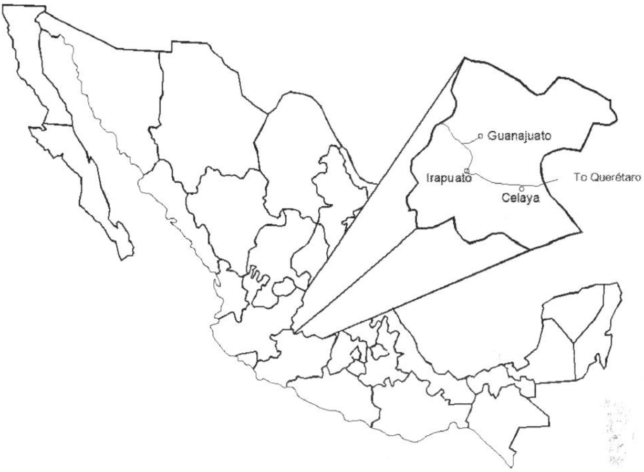

Figure 2. Study site

3.2. Analysis of the salt content

During the module drainage process (154 h), measurements of pH, temperature and electrical conductivity of water samples were made at defined time intervals (each hour the first 20 hours and subsequently increased to the range 2, 4, 6 and 8 h). The sensor used for measurement is a CONDUCTRONIC PC 18 sensor. The electrical conductivity at room temperature was recorded with it. However, in order to accurately quantify conductivity, it is important to consider a standard value of 25° C, which can be used to correct the values obtained. The correction factor used in accordance with [34] is 2-3 % for every Celsius degree that is measured under standard temperature. According with [34] (1964), the relationship between electrical conductivity and concentration is:

$$C = 640\ x\ EC \tag{23}$$

where C is the concentration given in $mg\,/\,l$ and EC the electrical conductivity given in $dS\,/\,m$ or $mmhos\,/\,cm$.

3.3. The hydrodynamic characteristic

To solve the Boussinesq equation, the van Genuchten model [25] for the water retention curve was used, along with a model of hydraulic conductivity of Fuentes [27] namely geometric mean model $\{K(\Theta)=K_s[1-(1-\Theta^{1/m})^{sm}]^2\}$ with the restriction $0<sm=1-2s\,/\,n<1$; where Θ is the effective saturation defined by $\Theta=(\theta-\theta_r)\,/\,(\theta_s-\theta_r)$.

3.4. The granulometric curve

The m and n form parameters from the water retention curve are obtained from the granulometric curve [35] adjusted with the equation $F(D)=[1+(D_g\,/\,D)^N]^{-M}$, where $F(D)$ is the cumulative frequency, based on the weight of the particles whose diameters are less than or equal to D; D_g is a characteristic parameter of particle size, M and N are two form empirical parameters. These parameters are rewritten as follows: $M=m$ and $N=[1\,/\,2(1-s)]n$.

3.5. Inverse problem

To evaluate the capacity of the numerical solution of the Advection-Dispersion Equation, the experimental information presented by [36] is used. The characteristics of the drainage module and the soil parameters used in the simulation are: $h_s=120\ cm$, $D_0=25\ cm$, $L=100\ cm$, $\phi=0.5695\ cm^3/cm^3$, and $s=0.7026$. The hydrodynamic characteristics used are those of van Genuchten and Fuentes [25,27]. The scale parameters $(\psi_d,\ K_s)$ are obtained from the inverse problem, using the experimental drained depth and the drained depth calculated with the numerical solution of the Boussinesq equation [10], given an error criterion between the previous and the new estimator $(1x10^{-12}\,cm)$, using a constant head test and fractal radiation condition with variable storage capacity and a nonlinear optimization algorithm [37]. The calculations were performed on a dual-core AMD Opteron machine with 2.6 GHz CPU and 8 GB RAM. The computational time required to solve the inverse problem was 5 h.

In order to model the salt concentration in the soil profile, with the numerical solution of the solute transport, the hydraulic parameters obtained from the previous analysis were used. In the numerical solution, the unknown parameter is the dispersivity coefficient (λ), which is estimated by minimizing the sum of squares errors between the salt concentration measured and the salt concentration calculated with the numerical solution over time, using a Levenberg-Marquardt [37], given an error criterion between the previous and the new estimator $(1x10^{-9}\,g\,/\,l)$. The initial condition is the sample initial, taken as a constant in all the system and radiation as the boundary condition applied in the drains.

4. Results and discussion

4.1. The granulometric curve

The adjusted parameters are shown in Table 1. Figure 3 shows the experimental granulometric curve and best fit is obtained with $D_g = 36.2993\ \mu m$ and $m = 0.3410$ with a root mean square error $RMSE = 0.1477$.

Model	Ajusted parameters			
	K_s	ψ_d	κ	$RMSE$
	(cm/h)	(cm)	(non-dimensional)	(cm)
Geometric mean model	1.5458	143.87	0.0616	0.2195

Table 1. Values of the adjusted parameters from the granulometric curve and the drained depth.

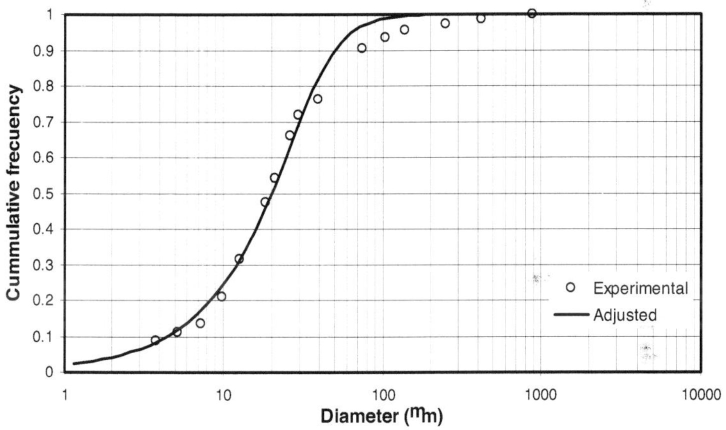

Figure 3. The experimental granulometric curve and adjusted with the model

4.2. The hydrodynamic characteristic

In order to obtain the values of ψ_d and K_s, the spatial and temporal increments used in all the simulation are $\Delta z = 0.0010\ cm$ and $\Delta t = 5x10^{-5}\ h$. Figure 4 shows the experimental drained depth and the drained depth calculated with the finite difference solution [10], using a storage capacity variable, fractal radiation condition in the drains and the geometric mean model. To linearize the boundary condition, one generalization of the conductance coefficient is optimized (κ) [8,10]. The residual volumetric water content is considered to be zero ($\theta_r = 0.0\ cm^3/cm^3$) [38].

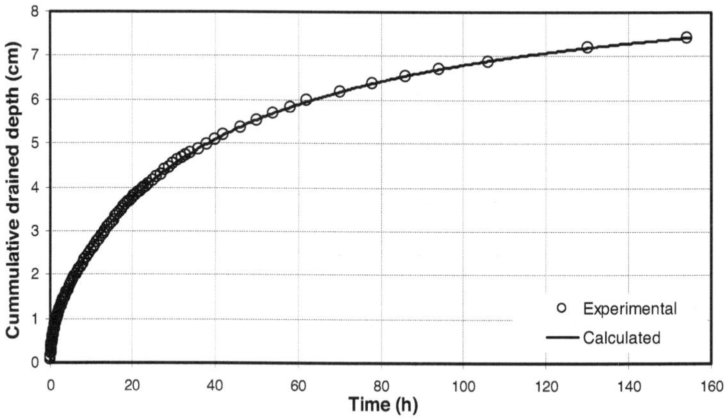

Figure 4. Comparison between the experimental drained depth and the calculated drained depth.

4.3. Analysis of the salt content

The EC data are shown in Figure 5 using a 2.5% like correction factor. Applying equation (23) to the data shown in Figure 5, we obtain the concentration in grams per liter (see Figure 6). The initial condition using in the numerical solution is the sample initial (C_{ini}=2.4g/l), taken as a constant in all the system and radiation as the boundary condition applied in the drains. The dispersivity value obtained is λ=91.80 cm, with RMSE = 0.1063 g/l between the experimental values and the values obtained from the numerical solution. The computational time required to solve the advection-dispersion model was 2.7 h. The dispersivity value found is only for this soil, because this value change with depth [39], increase with the flow rate and is a soil type function. This increase was explained by the activation of large pores at higher flow rates [40]. Figure 6 shows the experimental salt concentration evolution and the concentration obtained with the numerical solution.

Comparison shows that the salt concentration obtained with the numerical solution, according to RMSE, reproduce the experimental salt concentration. Figure 7 shows that in the short time, when the water flow increased, the salt concentration increases sharply, and in the long time tends to an asymptote, indicating that the system could not continue removing salts from the system. However, the value of the dispersivity obtained (λ=91.80 cm) overestimates the measured data in the long time. Second simulation was performed with the accumulated mass. To obtain the accumulated mass, it is necessary to obtain the solute flow, which is estimated by multiplying the water flow by the measured salt concentration in the time interval (see Figure 7). The cumulative solute mass is obtained by multiplying the solute flow by the time interval (see Figure 8).

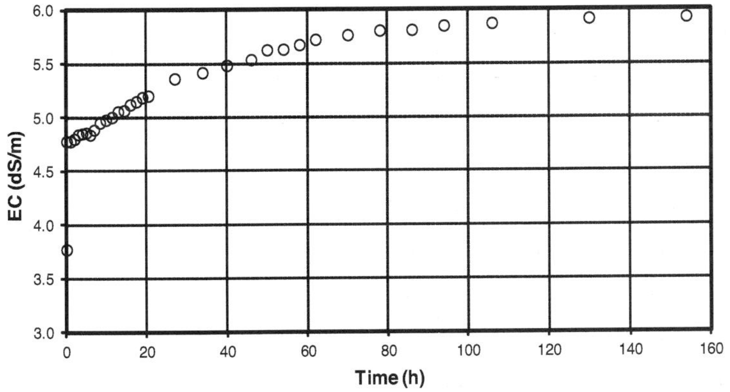

Figure 5. Evolution of the electrical conductivity of drainage water

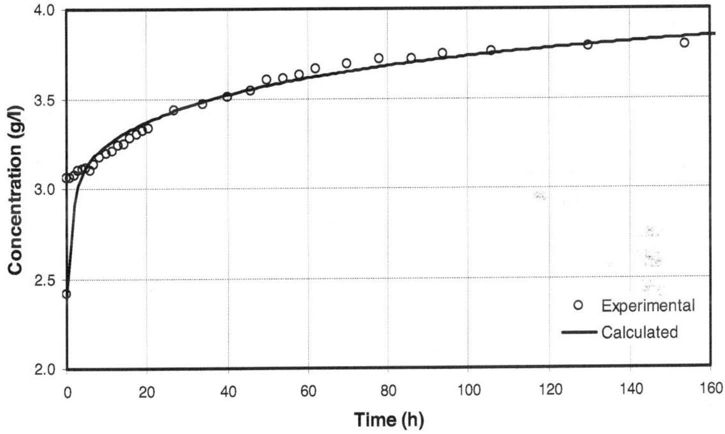

Figure 6. Comparison between the experimental and the calculated drainage water salt concentration with numerical solution

The results obtained with the numerical solution, the solute flow and cumulative mass evolution are shown in Figures 7 and 8, respectively, which demonstrate that the reproductions of the data were acceptable. The solute flow decreases rapidly, as seen in Figure 7 the concentration decreased 3.5 g/l after 20 hours. In the long time, the theoretical water flow and experimental water flow tends to be constant. Comparison showed that the solute flow and the cumulative mass evolution obtained with the numerical solution, according to RMSE, reproduce the experimental salt concentration. The RMSE values for estimating the solute flow and cumulative solute mass were $0.1842\ g/l$ and $0.1104\ g$, respectively. The dispersivity value obtained is $\lambda = 98.03\ cm$, with $RMSE = 0.1010\ g$ between the experimental values and the values obtained from the numerical solution. The dispersivity value for this new optimization (cumulative mass evolution) compared to the previous (salt concentration evolution) increases 6.2 cm.

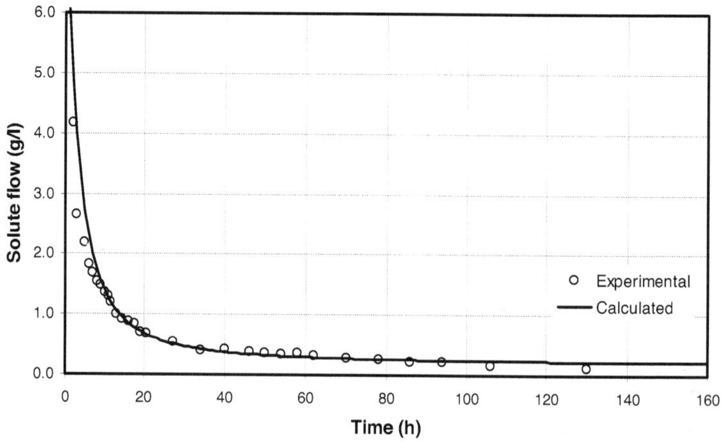

Figure 7. Solute flow (g/h) in the drainage system.

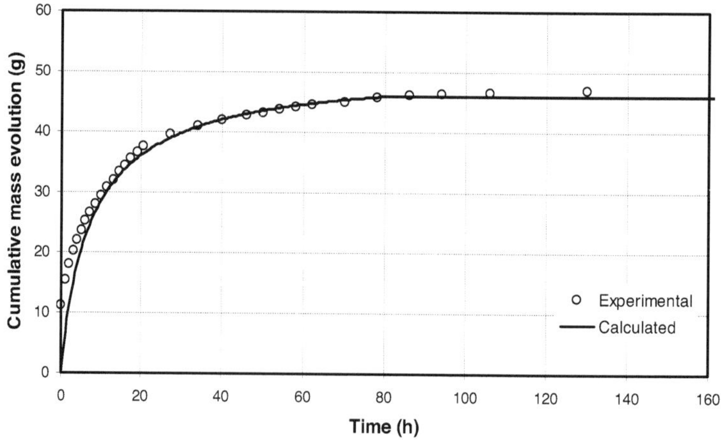

Figure 8. Cumulative mass evolution: experimental and obtained with the numerical solution.

4.4. Using the solution to simulate the leaching of saline soils

To recover saline soils it is necessary to apply irrigation so that the salts are transported to deeper horizons without harming the roots and are evacuated to other areas through the drainage channel. For purposes of illustrating the leaching of salts in the soil by applying the finite difference solution, we assumed a soil with hydraulic and hydrodynamic characteristics previously found. The initial soil concentration was 10 dS / m and the problem was reduced to finding the number of irrigation that must be applied to carry a given concentration.

The final average concentration obtained in the profile at the end of the first simulation was the initial concentration in the system for the next simulation, and so on. Figure 9 shows the reduced concentration of salts in the soil profile based on an initial concentration. The values shown are an average concentration in the soil profile at 1-m depth. Depth of drains was assumed to be 2.0 m.

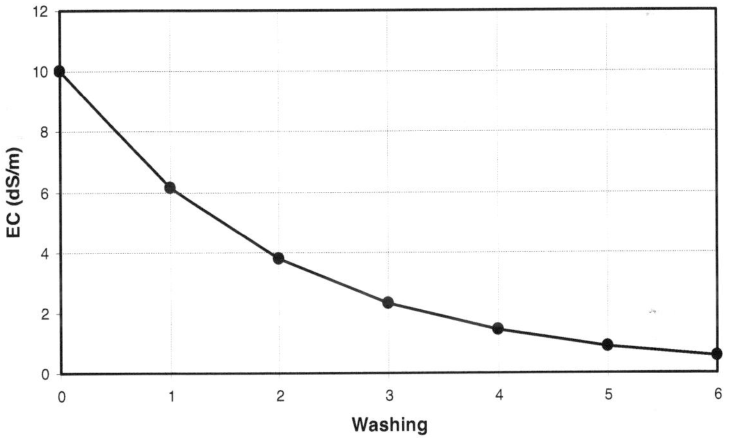

Figure 9. Evolution of the salt concentration in the soil by applying the leaching.

The simulations were performed with 5, 10, 15, 20 and 25 m of drains distances. It can be seen that the decrease in the concentration of salts in the soil profile is similar in all the separations between drains after applying 6 leachings. However, the time of drainage in each system was different. For example, with 5 days and 5 m of separation a decrease of the water table profile was more than one meter, while in the system with separation of 25 m decreased gives only a few centimeters (see Figure 10), other simulations was realized with 5, 10, 15, 20 and 25 m of drains distances, but the depth of drains was assumed to be 1.5 m (see Figure 11), therefore the time of drainage of the soil was a function of the distance between drains

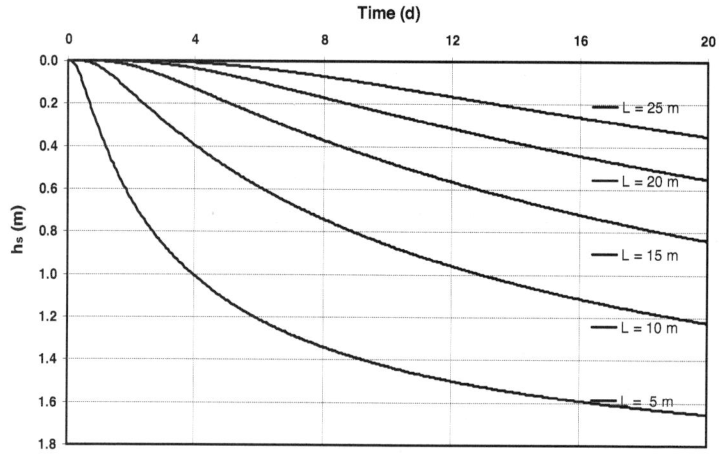

Figure 10. Decrease of the midpoint water table at different separations between drains under a drain depth of 2.00 m.

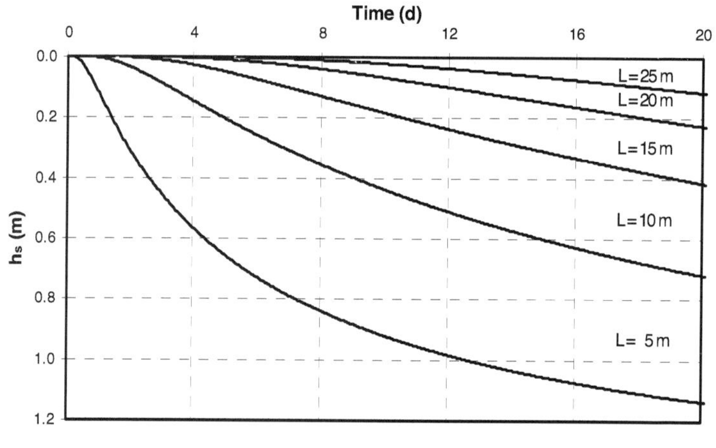

Figure 11. Decrease of the midpoint water table at different separations between drains under a drain depth of 1.50 m.

http://dx.doi.org/10.5772/59370

5. Conclusions

The irrigation in the arid and semi-arid regions to sustain agricultural production against the unpredictable of the rainfall have resulted in the double problem of salinity in many hectares of good agricultural land. Subsurface drainage systems are used to control the depth of the water table and to reduce or prevent soil salinity.

The advection-dispersion equation was solved in order to model the temporal evolution of the concentration of salts removed through an agricultural drainage system with the method of finite differences. The solution requires the values of the flow of water previously obtained from the solution of the Boussinesq equation. The hydrodynamic characteristic were obtained by the inverse problem from the depth drained.

The optimization of the accumulated mass which gave better results in terms of mean square error criterion between the theoretical and experimental values, since it is a property integrated in the time and concentration observed at specific levels. The solution presented coupled to the Boussinesq equation, satisfactorily reproduced the measured data, both in the short time where the change in concentration was high, as in long times where the concentration values tended to an asymptote. This asymptotic value of the concentration depended on the distance between drains of the drainage system.

Finally, the solution of differential equations of transfer processes of water and solute transport, and hydrodynamic characterization of the soil in an agricultural drainage system, will be a useful tool for designing new systems for the optimal development of crops according to their water needs and the degree of tolerance to salinity. In addition, this study can be help us for quantify crop yield reductions due to salinity on irrigation areas, in order to prevent future problems such as food shortages.

Author details

Carlos Chávez[1], Carlos Fuentes[1], Fernando Brambila[2] and Nami Morales-Durán[3]

*Address all correspondence to: chagcarlos@gmail.com

1 University of Queretaro, Faculty of Engineering, C.U. Cerro de las Campanas, Querétaro, México

2 National Autonomous University of Mexico, Faculty of Sciences, Ciudad Universitaria, México

3 University of Queretaro, Faculty of Natural Sciences, Avenida de las Ciencias S/N. Col. Juriquilla, Querétaro, México

References

[1] Ritzema HP., Satyanarayama TV., Raman S., Boonstra J. Subsurface Drainage to Combat Waterlogging and Salinity in Irrigated Lands in India: Lessons Learned in Farmers' Fields. Agricultural Water Management 2008; 95 (3) 179-189.

[2] Mousavil SF., Mostafazadeh-Fard B., Farkhondeh A., Feizi M. Effects of Deficit Irrigation with Saline Water on Yield, Fruit Quality and Water Use Efficiency of Cantaloupe in an Arid Region. Journal of Agricultural Science Technology 2009; 11 469-479.

[3] Ruiz-Cerda E., Aldaco NR., Montemayor TJ., Fortis HM., Olague RJ., Villagómez GJ. 2007. Aprovechamiento y mejoramiento de un suelo salino mediante vermicomposta. Tecnologías Pecuarias en Mexico 2007; 45 (1) 19-24.

[4] CONAGUA. Statistics on Water in Mexico (in Spanish). México. D.F; 2010.

[5] IMTA. Manual de Diseño e Instalación de Drenaje Parcelario en Zonas áridas y Semiáridas bajo Riego. México; 1998.

[6] Boussinesq J. 1904. Recherches The´oriques sur L'e´coulement des Nappes d'eau Infiltre´es Dans le Sol et Sur le De´bit des Sources. J. Math. Pure. Appl., 1904; 10 5-78.

[7] Verhoest N., Pauwels V., Troch P., de Troch F. Analytical Solution for Transient Water Table Heights and Outflows from Inclined Ditch-Drained Terrains. J. Irrig. Drain Eng., 2002; 128(6) 358–364.

[8] Zavala M., Fuentes C., Saucedo H. Nonlinear Radiation in the Boussinesq Equation of Agricultural Drainage. J. Hydrol. 2007; 332(3) 374-380.

[9] Singh S., Ghosh NC., Pandey RP., GalkateRV., Thomas T., Jaiswal RK. Numerical Solution of 1D Boussinesq Equation for Water Table Fluctuation between Drains in Response to Recharge and ET in A Sloping Aquifer. Int. J. Eco. Econ. Stat. 2009; 14(9) 45-54.

[10] Chávez C., Fuentes C., Zataráin F., Zavala M. Finite Difference Solution of the Boussinesq Equation with Variable Drainable Porosity and Fractal Radiation Boundary Condition. Agrociencia 2011; 45(8) 911-927.

[11] Taylor GI. The Disperson of Matter in Turbulent Flow Through a Pipe. Proc. R. Soc. London, Ser. A. 1954; 223 446–48.

[12] Elder JW. The Dispersion of Marked Fluid in Turbulent Shear Flow. J. Fluid Mech., 1959; 5 544-560.

[13] Fischer HB. The Mechanics of Dispersion in Natural Streams. J. Hydraul. Div., Am. Soc. Civ. Eng., 1967; 93(6) 187-216.

[14] Holly FM. Two Dimensional Mass Dispersion in Rivers. Hydrologic papers, Colorado State University Press, Fort Collins, Colorado, 1975; pp. 78.

[15] Holly FM. Dispersion in Rivers and Coastal Waters, 1, Physical Principles and Dispersion Equations. In: Novak, P. (ed). Developments in Hydraulic Engineering, 3. Elsevier, New York. 1985; Chap. 1: 1-38.

[16] Rutherford JC. River mixing, Wiley, New York; 1994.

[17] Mirabzadeh M., Mohammadi K. A Dynamic Programming Solution to Solute Transport and Dispersion Equations in Groundwater. J. Agric. Sci. Technol. 2006; 8 233-241.

[18] van Dam JC., Huygen J., Wesseling JG., Feddes RA., Kabat P., van Valsum PEV., Groenendijk P., van Diepen CA. Theory of SWAP, Version 2.0. Simulation of Water Flow, Solute Transport and Plant Growth in the Soil- Water-Atmosphere- Plant Environment, Department Water Resources, WAU, Report 71, DLO Winand Staring Centre, Technical Document 45, Wageningen; 1997.

[19] Simunek J., Sejna M., van Genuchten MTh. The HYDRUS-1D software package for simulating the movement of water, heat, and multiple solutes in variably saturated media, version 2.0, United States Salinity Laboratory, USDA-ARS, Riverside, Calif; 1998.

[20] Simunek J., van Genuchten MTh., Sejna M., Toride N., Leij FJ. The STANMOD Computer Software for Evaluating Solute Transport in Porous Media Using Analytical Solutions of Convection-Dispersion Equation, Versions 1.0 and 2.0, IGWMC – TPS-71, International Ground Water Modeling Center, Colorado School of Mines: Golden; 1999.

[21] Fayer MJ. UNSAT-H Version 3.0: Unsaturated Soil Water and Heat Flow Model. Theory, User Manual, and Examples. Pacific Northwest National Laboratory 13249; 2000, USA, 184 pp.

[22] Jansson PE., and Karlberg L. Coupled Heat and Mass Transfer Model for Soil-Plant-Atmosphere Systems, Royal Institute of Technology, Department of Civil and Environmental Engineering: Stockholm; 2001.

[23] Richards LA. Capillary conduction of liquids through porous mediums. Physics 1931; 1 318-333.

[24] Fuentes C., Zavala M., Saucedo H. Relationship between the Storage Coefficient and the Soil-Water Retention Curve in Subsurface Agricultural Drainage Systems: Water Table Drawdown. J. Irrig. Drain. Eng., 2009; 135(3) 279-285.

[25] van Genuchten MTh. A Closed-Form Equation for Predicting the Hydraulic Conductivity of the Unsaturated Soils. Soil Sci. Soc. Amer. J. 1980; 44 892-898.

[26] Carslaw HS., Jaeger JC. Conduction of Heat in Solids. Oxford University Press, Oxford; 1959.

[27] Fuentes C., Brambila F., Vauclin M., Parlange J-Y., Haverkamp R. Fractal modeling of Hydraulic Conductivity in Non-Saturated soils. Hydraul. Eng. México 2001; 16(2) 119-137.

[28] Abassi F., Simunek J., van Genuchten MTh., Feyen J., Adamsen FJ., Hunsaker DJ., Strelkoff TS., Shouse P. Overland Flow and Solute Transport: Model Development and Field-Data Analysis. J. Irrig. Drain. Eng. 2003; 129(2) 71–81.

[29] Zerihun D., Furman A., Warrick AW,. Sánchez CA. Coupled Surface-Subsurface Solute Transport Model for Irrigation Borders and Basin. I. Model Development. J. Irri. Drain. Eng. 2005; 131(3) 396-406.

[30] Simunek J. Models of Water Flow and Solute Transport in the Unsaturated Zone. Encyclopedia of Hydrological Sciences. Edited by M G Anderson; John Wiley & Sons, Ltd., Chichester, England, 2005; 1171-1180

[31] Zataráin F., Fuentes C., Palacios VOL., Mercado E., Brambila F., Villanueva N. Modelación del Transporte de Agua y Solutos en el Suelo (In Spanish). Agrociencia 1998; 32(4) 373-383.

[32] Freund RW., Hoppe RW. Stoer/Bulirsch: Numerische Mathematik 1. Springer-Lehrbuch, Germany; 2007.

[33] Conte SD., De Boor C. Elementary Numerical Analysis: An Algorithmic Approach. McGraw-Hill, New York; 1980.

[34] Villareal E., Bello S. The concentration and electrical conductivity in aqueous solutions of electrolytes. Rev. Mex. Fis. 1964; 13(2) 55-74.

[35] Fuentes C. Approche Fractale des Transferts Hydriques Dans les Sols No-saturés. Tesis de Doctorado, Universidad Joseph Fourier de Grenoble, Francia; 1992.

[36] Chávez C. Solución numérica de las ecuaciones de transferencia de agua y solutos en riego y drenaje. Dr. in Eng. Thesis, Universidad Autónoma de Querétaro (In spanish). México; 2010.

[37] Marquardt DW. An Algorithm for Least-Squares Estimation of Nonlinear Parameters. SIAM J. Appl. Math. 1963; 11 431-441.

[38] Haverkamp R., Leij FJ., Fuentes C., Sciortino A., Ross PJ. Soil Water Retention: I. Introduction of a Shape Index. Soil Sci. Soc. Am. J. 2005; 69 1881-1890.

[39] Simunek J., van Genuchten MTh. Using the HYDRUS-1D and HYDRUS-2D Codes for Estimating Unsaturated Soil Hydraulic and Solute Transport Parameters. pp. 1523–1536. In M.Th. van Genuchten. Simunek, J. and "Sejna, M. (ed.) Characterization and measurement of the hydraulic properties of unsaturated porous media. University of California, Riverside, CA; 1999.

[40] Feyen J., Jacques D., Timmerman A., Vanderborght J. Modelling Water Flow and Solute Transport in Heterogeneous Soils: A Review of Recent Approaches. J. Agric. Eng. Res. 1998; 70 231-256.

Chapter 5

Seeds of Change — Plant Genetic Resources and People's Livelihoods

Thomas Hilger, Iris Lewandowski, Bastian Winkler,
Barbara Ramsperger, Paulo Kageyama and
Carlos Colombo

Additional information is available at the end of the chapter

http://dx.doi.org/10.5772/59984

1. Introduction

Since the dawn of mankind people explore the cornucopia of the plant kingdom to obtain food, feed, fiber, and fuel. Worldwide, the total number of higher plants is estimated to be 270,000 species. About 43% of them are crops, cultivated plants, and their wild relatives, nowadays being classified as plant genetic resources for food and agriculture. Throughout the history only 7000 species representing almost 2.5% of the total were cultivated by mankind in the one or other way [1].

Mankind has largely benefited from modern agriculture. Today, agriculture provides food, feed, fuel, and fiber for more than 7,000 million people whereas a hunter-gatherer lifestyle supported around 4 million people only. In the past four decades, global cereal production almost doubled. This strong increase was based on greater use of inputs such as fertilizer, water, pesticides, new crop varieties, and other technologies provided by the 'Green Revolution'. This helped to diminish hunger, improve nutrition, and spare natural ecosystems from conversion to cropland. By 2050, the population is expected to have increased by 50%. Increasing and sustaining food production for a growing world population will, therefore, be a major challenge in future; however achieving this without compromising environmental integrity and public health is even more challenging due to changing habits in food consumption in parts of the world, a rising demand for biofuel, diminishing returns to fertilizer, and an increasing water demand [2].

Although new crop varieties substantially contributed to increase crop yields, specialized plant breeding has led to a strong dependence on few crops only along with erosion of plant genetic resources. For thousands of years farmers have been domesticating plant species thereby developing a wide range of crop varieties adapted to specific needs and environmental conditions [3]. Over the past 100 years the importance of many crop species has decreased strongly and many adapted crop varieties diminished. The private agricultural sector increased significantly in both developed and developing countries during the past 20 years; however, the main focus of its interest has been high-value products, such as maize, wheat, rice, oil crops, pulse crops and vegetable crops [4]. Today, 30 crops provide 95% of human food energy needs, 12 crops together with five animal species deliver 75 % of the world's food today, and three of which, i.e. rice, wheat, and maize, are responsible for more than 60% of our energy intake [5]. Almost 90% of global vegetable oil is produced by four crops only; i.e. oil palm, soybean, oil seed rape and sunflower [6]. Nowadays, our food security depends on a tiny number of crops only; it is therefore essential to sustain a high genetic crop diversity for both coping with increasing environmental stresses and facilitating farmers and researchers with options to breed cultivars adapted to less favourable conditions, such as salinity, poor soils and extreme weather events and that can resist biotic stresses, such as pests and diseases [4].

Many studies and reports discuss the importance of plant genetic resources for crop production in view of climate change and their key role in adapting to adverse climatic conditions and, hence, for food security. Important is that underutilized or minor crops often harbour high levels of genetic diversity being maintained on-farm in small-scale farming systems; however they are relatively neglected by formal research and development strategies, including breeding programs. Results from Peru hosting a wealth of native agro-biodiversity including many underutilized crops indicated the potential of a breeding approach for indigenous Andean crops, based on a combination of evolutionary and participatory methods to reach a balance between yield improvement and maintenance of genetic diversity [7]. These authors also highlighted the importance for upscaling such activities but mentioned the need to address methodological, financial and institutional issues for further expansion. Such activities are particularly important for areas threatened by climate change and to safeguard local food production. A global analysis of climate risks for crops in 12 food-insecure regions indicated South Asia and Southern Africa as two regions that, without sufficient adaptation measures, are likely to suffer from negative impacts on several crops important to large food-insecure human populations [8]. Therefore, these areas need special attention in crop development and breeding for food security based on local seed systems.

Minor crops can be important at a local, regional and national level but are often neglected at an international level. They are staple foods, contributing to food supply in certain periods and to a nutritionally well-balanced diet but also supply raw materials such as oils, fibres and dyes, providing options for income generation. Plant genetic resources are sometimes also well adapted to marginal soil conditions, an important feature in face of climate change and increasing soil degradation in densely populated regions. Minor crops are neglected as their advantages are known only locally and due to lack of markets, infrastructure for processing, and international research activities. Their potential, however, is often only poorly addressed and the loss of plant genetic resources reduces current and future options for mitigation in the agricultural sector [3].

http://dx.doi.org/10.5772/59984

On this background, this chapter intends to reflect the importance of plant genetic resources for people's livelihood and the human impact on plant genetic resources in general, looks at opportunities how the use of genetic resources and introduction of new or minor crops can contribute to improve people's livelihood and discusses tools for impact assessment.

2. Plant genetic resources and people's livelihood

2.1. Global distribution of plant genetic resources

In general, the highest number of species of vascular plants is found in the humid tropics and subtropics [9]. The species number strongly declines from the tropics and subtropics towards the temperate and polar zone north and south of the equator, indicating that hotspots of biodiversity are mainly associated with warm and humid tropical conditions and pointing to the global importance of these zones. Biodiversity hotspots are also abundant in regions rich in orographic structure or high in geo-diversity such as mountains and coastal regions (Fig. 1). Biodiversity hotspots are also important in terms of agriculture as they often coincide with or are, at least, close to Vavilov's centres of origin and areas of agricultural development.

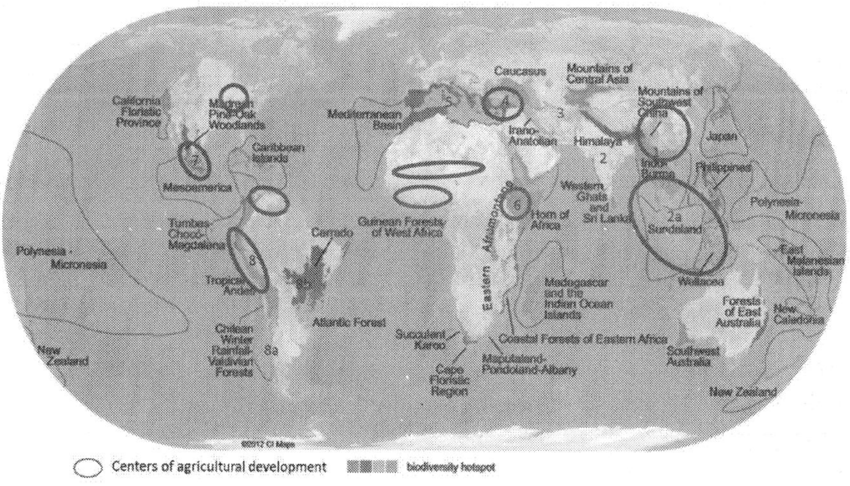

Figure 1. Biodiversity hotspots, Vavilov's centres of diversity, and centres of agricultural development. (Source: Image downloaded from Conservation International, 2013, modified)

In Africa, biodiversity hotspots are located in Madagascar and the Indian Ocean islands, the Congo basin, the eastern Arc Mountains and coastal forests, Guinean forests of West Africa, the Cape floristic province of South Africa, and the succulent Karoo. In Asia, hotspots are found in the Caucasus, the fertile half-moon region, Sundaland, Wallacea, the Philippines, the Indo-Burma region, the mountains of South-Central China, the Western Ghats of India, and Sri Lanka. Other important areas of biodiversity are located in the Mediterranean basin, South-western Australia, New Zealand, New Caledonia, and New Guinea and the Micronesian islands. Neotropical biodiversity hotspots are located in the tropical Andes, the Caribbean, Mesoamerica, the Atlantic forests, western Ecuador, the Chocó and Darién regions of northern South America, the Brazilian cerrados, central Chile, the upper Amazonia and Guyana shield, and the California floristic province. The estimated number of vascular plant species in Neotropical biodiversity hotspots ranges from 2-3000 up to more than 5000 species per 10,000 km^2 [9].

2.2. Plant genetic resources, minor crops and their potential in food, feed, fibre and fuel production

Diversity plays an important role as agricultural biodiversity is connected to nutrition and food security. Therefore, the primary justification for the conservation of plant genetic resources was given as their importance for breeding of improved varieties of crops for food, fuels and medicines [10]. In a rapidly changing world as today, conditions enhancing the adaptability and, hence, the resilience of family farms is crucial to their viability. Here, biodiversity can help farmers manage risks from new pests and diseases but can also lessen the effects of sudden natural disasters. Diversity allows natural adaptation to the environment which is vital in the face of climate change. Diversity also diminishes the risk of crop loss and contributes to productive, healthy farms [5].

Human infiltration into the tropical lowland rainforests of the Americas was late and slow as highlands and savannahs were easier to explore. This led to a lesser exploration of its biodiversity by mankind [11]. Nowadays, tropical rainforests of the Americas are esteemed as important biomes with regard to biodiversity, particularly when looking at industrial uses. Plant species originating from the Brazilian cerrado and Atlantic forest biomes provide promising bioactive secondary metabolites. These biologically active compounds are interesting for the pharmaceutical industry and present another option for a bio-products based income generation [12]. Palms originating from the Amazon lack key conditions for their sustainable management such as adoption of palm-climbing devices, not only for reducing wild palm felling but also for stimulating broader community-level conservation and efforts for their cultivation [13]. 'Improved extractivism' can be an appropriate way of growing wild plants, such as tucumã (*Astrocaryum* ssp.), a neotropical palm species rich in oil which can contribute to an increased farm income while increasing the economic value of disturbed areas in the central Amazon [14].

More than half the world's plant and animal species live in tropical forests. Hence, these biomes were often the cradle of food crops. The share of food crops derived from the tropics is high, estimated to be as much as 80%. They also provide genetic resources we will need if we are to produce new varieties, resistant to disease, insect pests, and climatic constrains, in the fight against world food shortages [15]. These biomes are also rich in non-timber forest products (NTFPs) important for developing new crops which include fruits and nuts, vegetables, fish and game, medicinal plants, resins, oils, essences, and a wide range of barks and fibres. Governments, conservation and development agencies and non-government organizations have encouraged their marketing and sale as a way of boosting income for poor people in the tropics and encouraging forest conservation. NTFPs are also of economic importance as basic raw material for various applications in industry, ranging from the development of new drugs against widespread diseases to bio-based compound composites [16].

The potential of Amazon's biodiversity in view of the automotive industry was underlined during a conference held at Belem, Para, Brazil in 1996 [17]. In particular, oil, fibre and resin producing plants are interesting in car manufacturing. Research results and experiences using natural fibres from the Amazon indicated their potential as reinforcement in composites for vehicles [18]. One of the tested fibres was curauá (*Ananas lucidus* syn. *A. erectifolius*), a hard-fibre producing bromeliad. Its fibre is well suited to substitute glass fibre in various parts of cars and trucks. Attempts to improve its productivity, harvest and processing, and to identify best agricultural practice were initiated [19-21]; however bridging the gap between needs of smallholders and industry improving the understanding of each other's bottlenecks was not possible due to a lack of both, use of participatory approaches and long-term commitment of the donor.

3. Human impact on plant genetic resources and options for their maintenance

There is rising awareness that habitat loss is associated with loss of genetic diversity; however the basic cause-effect relationships underlying the ecological roles of biodiversity are still poorly understood [22]. Major tropical wilderness areas show strong coincidence of biodiversity, people and concerns over watershed functions and are in a state of decrease or degradation as natural forests are converted into other land use classes and little area transforms back to natural forests or is reforested with plantations [23]. Various drivers are important for that. There is a high negative correlation of population density on species richness and tree diversity for tropical rainforests [24] but neither population nor poverty is the sole and major reason for land use and land cover change. It is driven by people's reaction to economic opportunities and constraints, created by local as well as national markets and policies. Additionally, global forces become main determinants as they intensify or weaken local factors [25].

Sustainability in agriculture suggests a focus on both improved understanding of the benefits of ecological and agronomic management, manipulation and redesign, and genotype improvements through the full range of modern biological approaches. A sustainable management of agro-ecosystems which includes aspects such as energy flows, nutrient cycling, human impacts and resilience may enhance reshaping of agriculture in landscapes. Sustainable cropping systems will influence food production, pesticide use and carbon stocks in a positive way. But the development of national and international policies supporting a stronger expansion of such systems in developed as well as developing countries still remains a major challenge [26]. Key drivers in the decline of biodiversity, its conservation and ecosystem services are the increased use of pesticides, herbicides and fertilizers, an increased homogeneity of landscapes associated with regional and farm-level specialization, drainage of water logged fields, loss of marginal and un-cropped habitat patches, and reduced fallow periods. Additionally, the intensification of agriculture has been fundamental to the degradation of ecosystem services and increases both, the production of greenhouse gases and a reduction of carbon sequestration [27].

Looking at the human impact on the earth system, nine planetary boundaries can be defined whereby the transgression of one or more of them is considered harmful due to abrupt environmental changes within continental- to planetary-scale systems. Three of them - climate change, the global nitrogen cycle and the rate of biodiversity loss - have already surpassed their threshold levels leading to consequences for global sustainability [28]. Strong negative impacts upon ecosystems are expected when the increasing annual global mean temperature rises above the pre-industrial mean by 2°C or more, especially in biodiversity hotspots [29].

The supply of non-timber forest products, timber and other services by forests resume a safety net function for rural populations providing both, income and jobs. Hence, fostering the role of forests in the political debate is essential, not only when looking at mitigation and adaption strategies for climate change but also for achieving sustainable developmental goals [30]. Experiences from Central Africa involving a wide range of stakeholders showed that focused research on priority sectors for poverty reduction are likely to raise public awareness of the forests' role and contribution to mitigate and to adapt to climate change at regional and national levels. A key for improving coherence and effectiveness of forest management policies is establishing the link between forests and climate change adaptation [30].

Plant genetic resources are essential for farmers to cope with future challenges, a feature emphasized by the FAO Commission on Genetic Resources for Food and Agriculture. New within-crop diversity will be needed to adapt to future conditions, and even new crops will be required under extreme conditions to reduce risks induced by climate change [4].

Despite progress in the past, crop wild relatives and underused species for food and agriculture need to be secured. The need for adapted germplasm is urgent and requires characterization, evaluation, and the availability of materials but gaps in *ex-situ* collections of selected crop gene pools are huge. The current focus on major crops leads to concerns due to a lack of *in-situ* and on-farm conservation. This neglects the importance of genetic, species, and ecosystems diversity. Currently, six million plant accessions are conserved in gene banks worldwide, representing a very limited number of species. Half of them are improved cultivars or breeders' lines. Only one third, however, represents landraces or old cultivars. About 15% refer to wild relatives of crop species and weeds. Minor crops and underutilized species are largely underrepresented in these collections, particularly primitive cultivars and wild relatives from the centres of origin's diversity and cultivation. Another obstacle is that only a third of all gene bank accessions have been fully characterized [1].

However, for many other crops, especially neglected or underutilized species and wild relatives of crops, comprehensive collections still do not exist and considerable gaps remain to be filled. There is also need for a better communication, collaboration and partnerships

http://dx.doi.org/10.5772/59984

among institutions dealing with the management of plant genetic resources, from conservation to plant breeding and seed systems [4].

Since farmers know best which materials meet their needs and are enthusiastic seekers of new varieties, "participatory plant breeding" represents a promising approach to enhancing agro-biodiversity [3]. Their participation in the whole process from the selection of plant genetic material up to development of cropping systems would improve and help to meet their needs, while also sustaining food security and alleviating poverty. Plant breeding of major crops, however, often lacks this participation, neglecting marginal site conditions and fostering high input demanding improved varieties.

Genetic diversity is essential for improving crops already in use but also for developing potential novel crops. The successful use of plant genetic resources and their sustained production depend to a large degree on access to genetically diverse germplasm. Free international exchange of germplasm will contribute significantly to the worldwide development of new industrial crops [31]. This is another challenge which needs to be solved in an equitable and fair way. Often indigenous knowledge is highly esteemed by people but once it comes to payment for such services people are reluctant to do so. Contracts for compensating creativity of farmers and a framework for rewarding grass root creativity and innovation are possible solutions [32], while consumers' and traders' responsibility is fostering fair trade, supporting local people in their production of food, feed, fibre and fuel [33]. Four types of collaboration are considered vital for using plant genetic resources in a sustainable way: (i) a national regulatory framework for biodiversity prospecting, (ii) the development of infrastructure and technology; (iii) formal contractual relationships among biodiversity's sources, intermediaries and final users; and (iv) the moving of research and development into the source country so as to contribute to its gross national product [16].

There are knowledge gaps with regard to the world's ability to match both a bio-based energy production and maintaining food security at the same time [34]. The main challenge is the competition for arable land and limited fresh water resources associated with a fast growing demand for food, feed, fibre, and fuel worldwide. A solution requires higher crop yields and improving resource-use efficiencies, especially that of nitrogen, and water productivity in production systems over the next two or three decades which is only possible with high external inputs. Cropping systems adapted to climate change and improved stress tolerance of crops are key issues in this context. This can be achieved by genetic improvement of crops and establishing sustainable cropping systems in diverse environments. Integrated assessments of their impact on resources, environments, and people's welfare can help identifying management options, species and varieties well adopted or most appropriate for specific environments. This will largely depend on the added value regarding specific ambitions, i.e. food, feed, fibre or fuel [34].

Energy wealth in Latin America has so far contributed little to overcome poverty and foster development. The sometimes considerable oil and gas earnings are not used to improve general welfare but skimmed off by the elite. A major problem is that both, energy reserves and energy utilization, are unevenly distributed in Latin America [35]. Using bioenergy is an option for an improved participation of rural communities if political leaders set the right frame.

There is lack of proper research, training, and socio-economic information to produce biofuels in a sustainable way. Hence, research in agriculture has to set a focus on improved crop selection based on the local situation and on management options including cultivation, management of pests and diseases, mechanisation, and harvesting. Furthermore, it is necessary to adapt cropping systems to local soil conditions and use by-products of biofuel crops to increase the efficiency of nutrient use and decrease negative influences on the environment. This is essential for the sustainable production of bio-fuel crops [36].

There are two main approaches to conserve plant genetic resources:

- *In-situ* conservation which is the protection of biological resources in their native environments and within naturally established and evolving populations, e.g. networks of protected areas which are ecologically representative of the forest types present on the landscape; sustainable forest management practices, ensuring that harvesting practices are genetically sustainable and ecologically compatible with the natural regeneration of target species, maintaining locally adapted gene pools and their genetic diversity *in situ*, and conforming with the requirements of other forest-dependent species that affect forest regeneration and health.
- *Ex-situ* conservation, particularly where *in-situ* conservation cannot be practiced or will not be sufficient to ensure adequate protection for genetic resources including germplasm banks and common garden archives, seed banks, tissue and cell cultures, cryopreservation, and DNA banks [37].

Both approaches can also be applied to conserve agro-biodiversity where many land-races have vanished since the green revolution in the 1960ies. However, effectively conserving wild biodiversity in agricultural landscapes will require increased research, policy coordination, and strategic support to agricultural communities and conservationists [38]. Research needs to address open questions regarding the minimum size and level of area connectivity required to conserve biodiversity *in-situ* at landscape level. *Ex-situ* conservation is biased by human decisions. But where shall we set priorities for sampling? Who pays for the costs of collecting and sustaining such kind of environmental services? Can value be added by exploring options for development of bio-products? Identifying common priorities in shared natural resource systems, however, is a major step in sharing a common responsibility in addressing climate change and associated problems.

4. Examples for the potential of introducing underused genetic resources

Globally, we are facing a pressure on vegetable oil markets with an increasing demand for food, fuel and chemical applications of vegetable oils on the one hand, and a limited potential for a sustainable extension of vegetable oil production on the other hand. The increasing demand for vegetable oils has led to the steep increase of oil palm production area in South-East Asia, nearly all of it on the cost of rainforest area and dramatic losses of biodiversity. In Europe the production of rapeseed oil increased strongly, too, on the expense of high inputs of agrochemicals. Rapeseed is one of the most demanding crops, requiring high inputs of nitrogen fertilizer and up to a dozen of applications of pesticides and insecticides. A further

http://dx.doi.org/10.5772/59984

extension of rapeseed in Europe is limited by the availability of suitable land and narrow crop sequences.

All over the world, we find various activities on testing and promoting plant genetic resources for vegetable oil and biofuel production. For this purpose especially plant species which do not compete with food crops growing on less fertile land are of particular interest. So far, real success stories, if at all, are rare. *Jatropha curcas*, a species endemic to the Brazilian cerrados, nowadays widely spread in the tropical zone received strong public attention in the past but never fulfilled people's partly exaggerated expectations [39]. Many studies on this species are very enthusiastic on reclaiming wasteland while simultaneously producing high oil yields. This often contributes to hype these species, although sound fundamental research is lacking to backup and foster farmers' adaptation [40].

Jatropha has gained international recognition as feedstock for bio-diesel production in the early 1980s. Its properties convinced investors, policy makers and clean development mechanism (CDM) project developers to consider it as a promising substitute for fossil fuel. Its toxic compounds exclude jatropha from human consumption. The same is true for the protein-rich press cake remaining after oil extraction which, hence, cannot be used as animal feedstuff. Jatropha grows on poor soil but yields are also poor under such conditions. This species is open-pollinating which hampers selection of specific lines for developing non-toxic varieties by breeding. Another advantage, often mentioned when arguing for this species is that jatropha is drought-resistant and well adopted to erratic rainfall conditions. Propagated by seeds it develops a deep rooting system. This is not the case when propagated by woody stem cuttings. Then, for higher yields sufficient water needs to be supplied in semi-arid or sub-humid regions with erratic rainfall conditions, particularly during early growth [39].

Jatropha, however, is partly still a wild plant of which basic agronomic properties are not fully understood, while environmental effects have not been investigated yet. Main knowledge gaps are found in the cultivation of the crop, for both a description of best practice as for describing the potential environmental risks or benefits. Therefore, fueling the jatropha bio-diesel hype has to be handled with care, unless the before mentioned knowledge gaps are closed by sound research [40].

The process of introduction of jatropha was characterized by top-down approaches, often neglecting the needs and involvement of local farmers. This often led to non-acceptance or even resistance against jatropha plantations [41]. Another obstacle is the fact that jatropha products are non-edible. Many farmers that have been establishing jatropha in Africa, partially on land previously used for food production, and that did not find a market or processing facility for the jatropha nuts, cannot use the products in case food is needed. This questions the approach of planting crops that deliver non-edible products because there is no flexibility in use deciding for either food or fuel use.

Activities in the biofuel sector are also driven by external forces, e.g. environmental concerns or a growing worldwide demand for biofuel, which generates political action. Brazil's government initiated a national bio-diesel policy, promoting feedstock supply from family farms. Especially in semi-arid regions, farmers have been encouraged to grow castor beans; however farmers' uptake of improved varieties was poor as the majority of farmers face great challenges associated with limited market access, top-down trading conditions, and lack of farmers' association fostering their market position. A stronger policy impact could be achieved by promoting bio-diesel crops that have alternative markets and fit more easily into the current farming system, reducing trade-offs with current crop activities and allowing synergies between fuel and feed production. Better enforcement of resource providing contracts is critical to avoid default and to alleviate labour and land constraints, thereby improving farmers' ability to engage in bio-diesel crop production [42].

In this context, another example endemic to the neotropics - the oil-producing macaw palm (*Acrocomia aculeata*) - has to be mentioned. Macaw palm recently gained economic importance in Paraguay and Brazil. In contrast to the African oil palm (*Elaeis guineensis*), it is adapted to a much wider range of environmental conditions which allows its production outside of the humid tropical zone, reducing negative impact on tropical rain forests. Another advantage of macaw palm is that it does not contain toxic compounds. The palm is a non-domesticated species with a high yield potential of an estimated 2.5 to 10.9 tons oil per hectare and year [43-45] and a life time of 70 years [46]. It grows well under various soil and weather conditions, naturally occurring in tropical and subtropical environments from southern Mexico to northern Paraguay and Argentina [46, 47]. It is often found on degraded grasslands as single trees, providing some extra feed to cattle which eat both, the fruits and leaves. The palms sustain longer periods without rain, and dry periods may last up to several months. Macaw palm fruits have a wide range of market opportunities with local and international perspectives as they are able to provide food, feed, fibre, and fuel (Fig. 2) [45]. The production and use of macaw palm can, therefore, provide a good example for a bio-economy crop that can fulfil food and fuel demands at the same time. Macaw palms growing in the Brazilian cerrados show a huge variability in biomass production and oil yield within and across various sites which highlights the importance of protecting biodiversity hotspots as source of future crops and in view of their domestication potential [48].

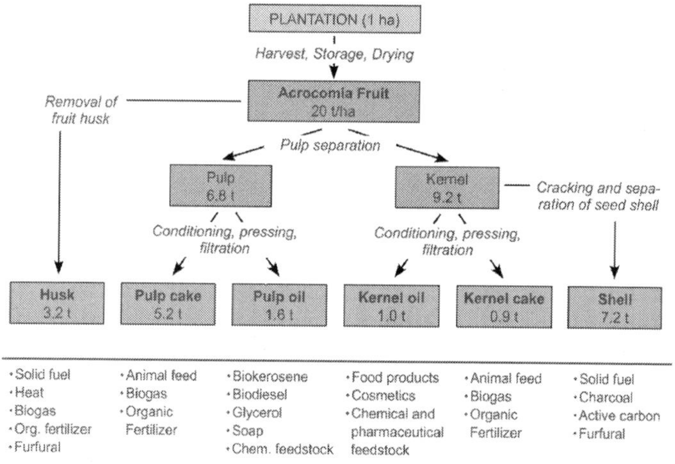

Figure 2. Processing, dry matter yield fractions and uses of macaw palm products [45]

http://dx.doi.org/10.5772/59984

5. Tools for the evaluation of impact assessment

The negative experiences of the large scale and partially forced introduction of the new crop jatropha [43] have shown that the introduction of a novel crop shall be guided by an *ex-ante* assessment of its ecological, economic and social impacts, including questions of local likelihood of acceptance.

Newcomers such as jatropha often lack proper long-term research on feasibility, trade-offs and environmental consequences, contributing to a better acceptance as well as public and private sector commitments for understanding needs of rural communities [39]. Macaw palms adapted to a wide range of environments are naturally occurring from Central America down to the north of Argentina and Paraguay [45], hence having potential for being cropped in many areas of South America or even outside of this continent, e.g. in Africa or Asia. Out-scaling of promising novel plants, however, also bears risks and requires an approach looking at all aspects of production from selection of genetic material and propagation, testing of cropping systems and crop management options to harvesting, transport, storage, and processing as well as considering development of new products. This needs an analysis of the entire value chain.

An *ex-ante* look at ecological and socio-economic aspects of macaw palm production or other novel crops allows identification of potential benefits, constraints and risks. Modelling is one approach in the portfolio of tools and techniques available to unravel dynamics of land use and their impact on the ecosystem associated with introduction of novel or alien species. Land use systems research addresses issues such as agricultural policy making, land use planning and integrated water management and involves for this purpose multiple stakeholders with various potential roles. Models are appreciated for both, their characteristic system research features and their integrative capacity [49]. Land use change models are tools for understanding and explaining causes and consequences of land use dynamics, The term land cover refers to the attributes of a part of the earth's land surface and immediate subsurface, including biota, soil, topography surface, groundwater and human structures. In that sense, modelling of land use changes provides insights into the extent and location of land use changes and its effects [50].

Especially important are arrangements regarding participation of stakeholders, and accountability in governance. Improving the ability of research programs to produce useful knowledge for sustainable development will require both greater and differentiated support for multiple forms of boundary work. Key issues are the use of knowledge for enlightenment, decision support, and negotiation support associated with boundaries between scientists and farmers, scientists and local policy-makers, and multiple knowledge sources and multiple users. Important determinants for their success are participation, accountability, and boundary objects to foster credibility and long-term success. Boundary objects are benchmark sites which allow studying human use of and impact on forest margins to gain knowledge for viable ways for a sustainable use of these sites. For decision support, joint creation of tangible products by scientists and farmers linking research with action, collaborative field trials, on-farm nurseries, and the production of training materials on effective land use practices are important. In terms of policy-makers, essential means are synoptic country reports, particularly when prepared as "policy briefs" on key issues and models focusing at regional scales. Moreover, it is important to note that context matters and challenges of boundary work need differentiation leading to strategies that follows context; however this is not possible without participation of all stakeholders [51].

Appropriate tools for that are:

- Participatory Landscape Analysis (PaLA),
- Rapid Carbon Stock Appraisal (RaCSA), and
- Integrated Renewable Energy Potential Assessment (IREPA).

PaLA was developed by the World Agroforestry Centre for agro-ecological analysis [52]. It captures local knowledge at relevant temporal and spatial scales, and provides insights on farmers' perception on the relationship between land use and landscape functioning: farmer's management options and the actual choices made, flows of water, sediment, nutrients and organisms, and internal filter functions that determine landscape functioning based on land use practices and interactions between landscape units. PaLA consists of the following eight steps [52]:

Step 1. Identification of ecological and administrative domains with clear boundaries.

Step 2. Sampling of representative stakeholders to be interviewed, using questionnaire and/or ranking methods. Criteria of representativeness are selected on the basis of specific project purposes.

Step 3. Formulation of the survey interdisciplinary group, planning and designing checklist and matching PRA tools.

Step 4. Making of a village sketch/model in order to identify the land use patterns and focus points in the landscape by using semi-structured interviews with male and female groups. The village sketch/model provides local names of area, distribution of land use, and main landscape features such as rivers, streams, mountains, roads.

Step 5. Transect walk are necessary to obtain an understanding of the soil-plant-water interactions along the landscape. Transects need to represent most of the land use types of the study area. The methods used are simultaneous transect walks and semi-structured interviews; delivered outputs are representative transects and sketches of the areas.

Step 6. A timeline for each land use type along transects or/and the fields located in the representative areas of the study catchment or village, is made to study land use changes over time, based on semi-structured interviews and timeline drawing.

Step 7. Feedback meeting in order to report findings to the farmers/stakeholders involved to get their feedback. The methods used are posters using visualising tools and group meetings.

Step 8. Data analysis: Qualitative data of each PRA tool, i.e. sketch transect, timeline, and secondary data is analyzed separately by the team. Thereafter, results are evaluated to identify landscape patterns and issues.

http://dx.doi.org/10.5772/59984

RaCSA also developed by the World Agroforestry Centre is a negotiation support tool that aims at providing reliable data on above and below ground carbon stocks in a defined landscape, its historical changes, the impact of on-going land-use change on projected emissions, and a framework for data generation on land-use options and their changes over time [52]. This approach assesses local ecological knowledge, explores its economic potential and uses carbon stocks as an indicator for the health or fertility status of soils. Furthermore, drivers of land use change and impact on environmental services such as biodiversity can be assessed. Simultaneously it provides knowledge on alternative land use options and mitigation strategies by means of mid- to long-term scenarios at landscape level using the Forest, Agroforest, Low value Landscape Or Wasteland (FALLOW) model. Finally, RaCSA was developed as negotiation support tool providing a basis for stakeholder discussions.

RaCSA consists of the following six steps [52]:

Step 1. Initial appraisal of landscape (see PaLA), focused on dynamics of tree cover.

Step 2. Explore Local Ecological Knowledge (LEK) and economics of local tree/forest management combined with a rapid household socio-economic survey.

Step 3. Plot-level C data of representative land cover units using an updated version of the ASB C_{stock} protocol provides time-averaged carbon stock data for above-ground vegetation and soils.

Step 4. Remote sensing and ground-truthing are used to provide spatial analysis of land cover change, based on a sufficiently sensitive 'legend'.

Step 5. The Public/Policy Ecological Knowledge (PEK) kit is used to obtain information on tree/forest management and existing spatial planning rules.

Step 6. Scenario studies of changes in C stocks and welfare through modelling land use and carbon stock dynamics in the landscape by using FALLOW.

FALLOW was developed by the World Agroforestry Centre for trade-off analysis [53, 54]. It allows a spatially explicit and dynamic modelling of land-use cover change (LUCC) in data-poor regions and merges bio-physical and socio-economic information to evaluate impacts of LUCC on food security, watershed functions, biodiversity, and carbon stocks. FALLOW was successfully applied in South-East Asia to assess local land use change dynamics without the need for long-term and data-intensive studies without the need for long-term and data-intensive studies based on farmers' knowledge [55], to explore livestock fodder options and their consequences for carbon stocks [56] and stakeholders' perceptions [57].

The introduction of new technologies, such as renewable energy technologies (RET) is comparable to the introduction of new plant species into local agricultural systems: new plant species as well as new technologies imply a change in the daily routine of local livelihoods. Traditionally, the planning of rural energy development projects took place in central government offices far away from rural communities [58]. The applied decision support tools aimed at the identification of the most efficient technology with the lowest costs [59]. The technologies selected in such 'top-down' approaches, were afterwards (involuntarily) imposed into rural communities [58]. Amigun *et al.* explored the community perspectives on the introduction of large-scale biodiesel production from canola (*Brassica napus*) and soybean (*Glycine max*) in South Africa [60]. The local population was overwhelmingly against the proposed biodiesel production. Their reasons for the rejection included a variety of especially social and environmental factors: land regarded as identity; competition with food security; distortion of the social community fabrics; doubts about the credibility of the developers and possible air and water pollution with respect to health risks of local population.

Several studies pointed out that the acceptance of RET depends on the complex interaction of social, institutional, environmental and techno-economic factors on a very local level [61, 62]. Smallholder agricultural systems are very diverse and therefore an assessment of these factors on individual basis is required that is based on public participation and pooled learning among the relevant stakeholders [60].

IREPA provides a people-centred, bottom-up approach for the assessment of the implementation potential of renewable energy technologies into smallholder agricultural systems [63]. This participatory approach explores the renewable resource base and the livelihoods of smallholder farmers to characterize the role of energy in the daily routine (social, institutional, environmental, technical, and economic factors) to select appropriate RET. The researcher acts as facilitator to guide the assessment while the local stakeholders become researchers who contribute knowledge and expertise. For that the IREPA approach comprises the following steps [63]:

- The assessment of local renewable resources based on a combination of statistical databases with global coverage and on-ground measurements for biogenic resources;
- The exploration of local social, institutional, environmental, technical and economic factors at household and community level by employing "participatory learning and action research methods" [64, 65].
- The combination of locally available resources and relevant factors to pre-select and design locally appropriate RETs;
- A participatory assessment of the impacts of these RETs on local livelihoods;
- A participatory identification of the most appropriate RET for implementation within a specific context using the multi-criteria decision analysis method "Analytical Hierarchy Process" (AHP) [66].

This participatory, bottom-up research structure provides a shift from the traditional top-down approaches to a holistic consideration of the local diversity. It aims to successfully induce changes into prevailing structures and behaviour patterns of smallholder agricultural systems in order to make sustainable use of the local natural resource base.

An important impact to the ecological and economic performance of land use systems is the productivity of these systems. To address this, the selected land-use systems and their performance can be modelled by using the **Wa**ter, **Nu**trient, **L**ight **C**apture in **A**groforestry **S**ystems model. This model deals with a wide range of agroforestry systems and annual single cropping systems with minimum parameter adjustments [67]. Hence, it is possible to explore the performance of various land-use systems under a wide range of management options and changing environmental conditions. In Brazil this model has already been used to assess the performance of sugar cane in agroforestry systems [68]. The model can be applied and adapted to various climate, soil, and cropping conditions [69, 70].

Agriculture is a major water consumer [71]. Furthermore, the semi-arid and arid areas of Brazil will suffer from a decrease of water resources due to climate change [72]. Brazil is a water rich country, for which the Amazon region is an example. But inappropriate land uses in Atlantic forest and cerrado ecoregions have led to a degradation of the soils which makes these resources scarcer in terms of biodiversity [73]. Different land use systems, small family farmers and industrial agriculture properties require different water quantities for their production processes. Therefore, water use efficiency will become an increasingly important sustainability indicator for land use systems. Water use in land use systems can be measured by evapotranspiration of crops. A water balance of the systems is an approach to quantify the water cycle. The water footprint is an indicator of water use, e.g. during agricultural processes, differentiating water use into three water categories: green, blue, and grey [74]. Green represents the volume of rainwater consumed during the production process and blue the natural run-off through groundwater and rivers minus environmental flow requirements, while grey is an indicator of freshwater pollution that can be associated with the production of a product over its full supply chain. The global water footprint in the period 1996-2005 was 9087 Gm^3 per year (74% green, 11% blue, and 15% grey) in which agricultural production contributed 92% [75].

We, therefore, propose linking and integrating the above mentioned methods once promising novel species or crops are identified to (i) develop and test land use options and model their agronomic performance; (ii) analyse, quantify and compare the ecological and socio-economic performance of land-use systems; and (iii) test options for developing new bioproducts. This procedure will help identifying optimised value chains, addresses trade-offs and consequences for environmental services, and looks at development of proper production systems for smallholder farmers. Using participatory approaches will foster farmers' participation and help identify options meeting their needs. It will also allow an exchange of ideas and information among all stakeholders.

6. Final consideration

More and more countries on all continents are developing bio-economy strategies striving for the sustainable use of renewable resources, especially biomass. Beginning with the development of bioenergy programs, bio-economic activities are growing rapidly in several countries and require an increasing supply of sustainably produced biomass. Here, the use of genetic resources to develop existing and new crops for a variety of applications in bio-based products can play an important role because bio-based products with new and improved properties also require a range of biomass properties. However, the development of bio-economies should not make the same mistakes that were observed in the development of modern bioenergies, such as transportation biofuels. The development of the modern bioenergy sector neglected the demands of smallholder farmers, who in the end did not benefit from the activities surrounding bioenergy but rather suffered the effects of land grabbing. Another much criticized effect of modern bioenergy development was the concentration on a few major crops only, such as maize or oil palm. Therefore we suggest that bio-economy strategies should ensure sustainable development of the biomass resource by:

a. *Ex-ante* assessment of the potential impacts of biomass production and supply systems

In section 5 of this chapter, various instruments were suggested for assessing the ecological and social impacts of biomass production systems which can be applied to *ex-ante* analysis and the planning of sustainable biomass production and supply systems.

b. Involving stakeholders and smallholder farmers

Acceptance of new technologies or crops and varieties is the pre-requisite for their implementation. Therefore their development should involve stakeholders, in particular smallholder farmers. Their involvement would not only improve the chances of implementation but also incorporate local or indigenous knowledge into developments.

c. Using genetic resources for developing new bio-economy crops or improved varieties

As discussed in this contribution there are many untapped genetic resources, most of them close to important agricultural centres. Demand for new crops and improved varieties for a bio-economy requires specific biomass properties on the one side, but also ecological requirements, such as nutrient- and water-use efficiency and stress resistance on the other. Therefore, in particular multi-purpose crops that integrate different land-use functions and biomass-use options appear most interesting for a future bio-economy. An example for such a multi-purpose crop was discussed in this contribution using the example of the macaw palm. This palm can grow under conditions of abiotic stresses, such as drought, and also contributes through its perennial character to soil carbon sequestration and erosion prevention. Its products are manifold allowing the integration of food, feed, fibre, and fuel production.

d. Development of sustainable biomass production concepts

The increasing demand for biomass leads to increasing pressure on land which can result in land-use changes, such as conversion from grassland to crop land. Recent findings from marginal grasslands show that increasing pressure on them can negatively influence ecosystem functioning, potentially compromising long-term production potentials. On the other hand, grassland communities in Europe suffer from mismanagement or under-management. In Europe many grasslands are no longer harvested due to the decreasing demand for roughage fodder. However, the maintenance of different grassland species requires cutting, but in regimes that are adapted to the ecological needs of grassland species. Therefore, biomass production concepts need to be developed that integrate production and ecological functions. Understanding the direct, indirect, and interactive effects of land-use changes on communities and ecosystems can help to better assess and balance such inherent trade-offs among multiple ecosystem functions [76].

There is need to prove such findings for other tropical environments such as rainforests on which a strong pressure lasts due to global interest on crops such as soybean and African oil

http://dx.doi.org/10.5772/59984

palm. The pressure on ecosystems resulting from the production of traded biomass, however, is highly variable between regions and products. Biomass consumption and trade are expected to surge over the next decades, suggesting a need to sustainably manage supply and demand of products of ecosystems on a global level [77].

There is need to find a way how to share and preserve biomes rich in species. Crops and other domesticated plants have a higher mobility than wild species and are already adapted to a wide range of environments due to anthropogenic influence in the past. Potentially new crops have undergone only little human selection if it all and are, hence, less well adapted to changes. In terms of gathered wild plants this may even be more severe. Recent studies indicate that under appropriate conditions, most native taxa may be sustainable within anthromes while at the same time increasing anthrome productivity in support of human populations [78]. The gaining economic interest on plant genetic resources for food and agriculture and their sustainable use will contribute to a better recognition and esteem of biomes hosting them. Proper rules for data and information exchange, for using plant genetic resources and indigenous knowledge are required. There is need for improving the south-north dialogue and creating the ground for south-south cooperation at international level to avoid further degradation, land grabbing and bio-piracy and foster local initiatives to safeguard locations rich in biodiversity and preserve land races *in-situ* on farmers' fields. As Robert Green Ingersoll, a British philosopher and ecologist of the 18th century, once said: "In nature there are neither rewards nor punishments – there are consequences" or in other words: Either we improve the conservation of plant genetic resources or we lose them forever without having had any opportunity to explore their potential.

Author details

Thomas Hilger[1*], Iris Lewandowski[2], Bastian Winkler[2], Barbara Ramsperger[3], Paulo Kageyama[4] and Carlos Colombo[5]

*Address all correspondence to: thomas.hilger@uni-hohenheim.de

1 Institute of Plant Production and Agroecology in the Tropics and Subtropics, Hohenheim University, Stuttgart, Germany

2 Institute of Crop Science, Hohenheim University, Stuttgart, Germany

3 Center for Agriculture in the Tropics and Subtropics, Hohenheim University, Stuttgart, Germany

4 Department of Forest Sciences, ESALQ, University of Sao Paulo, Piracicaba, Sao Paulo, Brazil

5 Instituto Agronômico de Campinas, Campinas, Sao Paulo, Brazil

References

[1] Hammer K, Arrowsmith N, Gladis T. Agrobiodiversity with emphasis on plant genetic resources. Naturwissenschaften 2003;90(6) 241-50.

[2] Tilman D, Cassman KG, Matson PA, Naylor R, Polasky S. Agricultural sustainability and intensive production practices. Nature 2002;418 671-677. http://www.nature.com/nature/journal/v418/n6898/pdf/nature01014.pdf (assessed September 26, 2014)

[3] Deutsche Gesellschaft für Internationale Zusammenarbeit (GIZ). Farmers as breeders – Participatory plant breeding. Issue Papers – People and Biodiversity. Working Paper. Eschborn, Germany: GTZ Imprint. 2009. star-www.giz.de/dokumente/bib/04-5104a1.pdf (assessed on January 7, 2009).

[4] Food and Agriculture Organisation (FAO). The Second Report on the State of the World's Plant Genetic Resources for Food and Agriculture. Rome: FAO. 2010. http://www.fao.org/docrep/013/i1500e/i1500e00.htm (assessed September 26th, 2014)

[5] Ann Tutwiler. Agricultural biodiversity: the foundation of resilient family farms. Rural 2014;21(2) 24-6. http://www.rural21.com/uploads/media/rural2014_02-S24-26.pdf (assessed September 26th, 2014).

[6] Carter C, Finley W, Fry J, Jackson D, Willis L. Palm oil markets and future supply. European Journal of Lipid Science and Technology 2007;109(4) 307-14.

[7] Galluzzi G, Estrada R, Apaza V, Gamarra M, Pérez Á, Gamarra G, et al. Participatory breeding in the Peruvian highlands: Opportunities and challenges for promoting conservation and sustainable use of underutilized crops. Renewable Agriculture and Food Systems. 2014; 1-10. DOI: 10.1017/S1742170514000179 (in press).

[8] Lobell DB, Burke MB, Tebaldi C, Mastrandrea MD, Falcon WP, Naylor RL. Prioritizing Climate Change Adaptation Needs for Food Security in 2030. Science 2008;319(5863) 607-610. http://www.sciencemag.org/content/319/5863/607.full.pdf (assessed September 26, 2014)

[9] Barthlott W, Lauer W, Placke A. Global Distribution of Species Diversity in Vascular Plants: Towards A World Map Of Phytodiversity. Erdkunde 1996;50 317-27.

[10] Ford-Lloyd B, Jackson M. Plant genetic resources: an introduction to their conservation and use. Cambridge: Cambridge University Press 1992.

[11] Harlan JR. Crops and Man, 2nd Edition. Madison, Wisconsin, USA: American Society of Agronomy; 1992.

[12] Bolzani VdS, Valli M, Pivatto M, Viegas C. Natural products from Brazilian biodiversity as a source of new models for medicinal chemistry. Pure and Applied Chemistry. 2012;84(9) 1837-46.

[13] Manzi M, Coomes OT. Managing Amazonian palms for community use: A case of aguaje palm (Mauritia flexuosa) in Peru. Forest Ecology and Management 2009;257(2) 510-17.

[14] Schroth G, Da Mota M, Lopes R, De Freitas A. Extractive use, management and in situ domestication of a weedy palm, Astrocaryum tucuma, in the central Amazon. Forest Ecology and Management 2004;202(1) 161-79.

[15] Newman A. Tropical Rainforest. New York: Facts on File Publisher. 1990

[16] Reid WV, Laird SA, Meyer CA, Gámez R, Sittenfeld A, Janzen DH, et al. Biodiversity prospecting: using genetic resources for sustainable development. Washington (DC): World Resources Institute; 1993.

[17] Panik F. The use of biodiversity and implications for industrial production. A third millennium for humanity. In: Leihner DE, Mitschein TA. (Eds.) A Third Millennium for Humanity? The Search for Paths of Sustainable Development. Frankfurt: Lang Publisher.1998. p.59-73.

[18] Kübler E. Use of Natural Fibres as Reinforcement in Composites for Vehicles: Research Results and Experiences. In: Leihner DE, Mitschein TA. (Eds.) A Third Millennium for Humanity? The Search for Paths of Sustainable Development. Frankfurt: Lang Publisher. 1998. p.393-402.

[19] Zah R, Hischier R, Leão A, Braun I. Curauá fibers in the automobile industry–a sustainability assessment. Journal of Cleaner Production 2007;15(11) 1032-40.

[20] Berger N. Agronomische Optimierung des Anbaus von Curauá (Ananas lucidus Miller) in der östlichen Amazonas Region von Brasilien. Weikersheim, Germany: Margraf; 2004.

[21] Behrens D. Curaua-Faser - eine Pflanzenfaser als Konstruktionswerkstoff? Untersuchungen zum Anbau der brasilianischen Bromelien-Art Curaua (Ananas lucidus Miller) und Prüfung der Blattfasern im Hinblick auf ihre technologische Nutzbarkeit. Berlin, Germany: Köster; 1999.

[22] Tomich TP, Thomas DE, van Noordwijk M. Environmental services and land use change in Southeast Asia: From recognition to regulation or reward? Agriculture, Ecosystems and Environment 2004;104(1) 229-244.

[23] Food and Agriculture Organisation (FAO). The status of forests: the Global Forest Resources Assessment 2000 Part II. Key Issues in the Forest Sector Today. Rom: FAO, 2001, http://www.fao.org/docrep/003/y0900e/y0900e05.htm (assessed September 26, 2014).

[24] Top N, Mizoue N, Ito S, Kai S, Nakao T, Ty S. Effects of population density on forest structure and species richness and diversity of trees in Kampong Thom Province, Cambodia. Biodiversity and conservation 2009;18(3) 717-38.

[25] Lambin EF, Turner BL, Geist HJ, Agbola SB, Angelsen A, Bruce JW, et al. The causes of land-use and land-cover change: moving beyond the myths. Global environmental change 2001;11(4) 261-9.

[26] Pretty J. Agricultural sustainability: concepts, principles and evidence. Philosophical Transactions of the Royal Society B: Biological Sciences 2008;363(1491) 447-65.

[27] Pretty J, Sutherland WJ, Ashby J, Auburn J, Baulcombe D, Bell M, et al. The top 100 questions of importance to the future of global agriculture. International journal of agricultural sustainability 2010;8(4) 219-36.

[28] Rockström J, Steffen W, Noone K, Persson Å, Chapin III FS, Lambin E, et al. Planetary boundaries: exploring the safe operating space for humanity. Ecology and Society 2009;14(2). http://www.ecologyandsociety.org/vol14/iss2/art32/ (assessed September 26, 2014)

[29] Warren R, Price J, Fischlin A, de la Nava Santos S, Midgley G. Increasing impacts of climate change upon ecosystems with increasing global mean temperature rise. Climatic Change 2011;106(2) 141-77.

[30] Sonwa DJ, Nkem JN, Idinoba ME, Bele MY, Jum C. Building regional priorities in forests for development and adaptation to climate change in the Congo Basin. Mitigation and Adaptation Strategies for Global Change 2012;17(4) 441-50.

[31] White G, Gardner J, Cook C. Biodiversity for industrial crop development in the United States. Industrial Crops and Products 1994;2(4) 259-72.

[32] Chand VS, Patel KK, Murali Krishna S, Gupta AK. Contracts for 'Compensating' Creativity: Framework for Rewarding Grassroot Creativity and Innovation. In: Leihner D. and MitscheinTA (Eds.) A Third Millennium for Humanity? The Search for Paths of Sustainable Development. Frankfurt: Lang Publisher. 1998. p201-217.

[33] Leatherman P. Fair Trade Organisations: Is This the Fairest Trade of All? In: Leihner DE. and Mitschein TA. (Eds.) A Third Millennium for Humanity? The Search for Paths of Sustainable Development. Frankfurt: Lang Publisher. 1998. p219-224.

[34] Spiertz J, Ewert F. Crop production and resource use to meet the growing demand for food, feed and fuel: opportunities and constraints. NJAS-Wageningen Journal of Life Sciences 2009;56(4) 281-300.

[35] Linkohr R. Latin America's energy policy between state and market. Internationale Politik und Gesellschaft 2006;4 105-119+191.

[36] Morgan K, Gilbert R, Helsel Z, Buacum L, Leon R, Perret J. White paper report from working groups attending the international conference on research and educational opportunities in bio-fuel crop production. Biomass and Bioenergy 2010;34(12) 1968-72.

[37] Rajora OP, Mosseler A. Challenges and opportunities for conservation of forest genetic resources. Euphytica. 2001;118(2):197-212.

[38] Scherr SJ, McNeely JA. Biodiversity conservation and agricultural sustainability: towards a new paradigm of 'ecoagriculture'landscapes. Philosophical Transactions of the Royal Society B: Biological Sciences. 2008;363(1491):477-94.

[39] Jongschaap R, Corré W, Bindraban P, Brandenburg W. Claims and Facts on Jatropha curcas L. Global Jatropha curcas evaluation, breeding and propagation programme. Wageningen, the Netherlands: Plant Research International BV, Report 158, 42 pp+ annexes. 2007.

[40] Achten W, Verchot L, Franken YJ, Mathijs E, Singh VP, Aerts R, Muys B. Jatropha bio-diesel production and use. Biomass and bioenergy 2008;32(12) 1063-84.

[41] Burnod P, Gingembre M, Andrianirina Ratsialonana R. Competition over Authority and Access: International Land Deals in Madagascar. Development and Change 2013;44 357-79.

[42] Leite JGDB, Bijman J, Giller K, Slingerland M. Biodiesel policy for family farms in Brazil: One-size-fits-all? Environmental Science and Policy 201327 195-205.

[43] Berton LHC, Azevedo JA, Filho CRL, Carvalho WJ, Siqueira RM, Coelho, and Colombo C A. Teor de óleo do fruto da macaúba (Acrocomia aculeata) nos Estados de São Paulo e Minas Gerais. Paper presented at 5th Congresso da Rede Brasileira de Tecnologia de Biodiesel. Held at Salvador de Bahia, Brazil, April 16-19, 2012. http://oleo.ufla.br/site/trabalhos/buscar-trabalhos/#!/codigo/80f50668946acd05603f935cb597db22 (assessed September 26, 2014)

[44] Pires TP, dos Santos Souza E, Kuki KN, Motoike SY. Ecophysiological traits of the macaw palm: a contribution towards the domestication of a novel oil crop. Industrial Crops and Products 2013;44 200-10.

[45] Poetsch J, Haupenthal D, Lewandowski I, Oberländer D, Hilger T. Acrocomia aculeata - a sustainable oil crop. Rural 21 2012;21(03) 41-4. http://www.rural21.com/uploads/media/rural2012_03-S41-44.pdf (assessed September 26, 2014)

[46] Agroenergías. "South American oil palm (Acrocomia aculeata Jacq.) as a new multipurpose crop." Acrocomia Solutions. http:// www.acrocomiasolutions.com/en/ (assessed September 15, 2014).

[47] Teles HdF, Pires LL, Garcia J, Rosa JQS, Farias JG, Naves RV. "Ambientes de ocorrência naural de macaúba." Pesquisa Agropecuária Tropical 2011;41 595-601.

[48] Ciconini G, Favaro S, Roscoe R, Miranda C, Tapeti C, Miyahira M. Biometry and oil contents of *Acrocomia aculeata* fruits from the Cerrados and Pantanal biomes in Mato Grosso do Sul, Brazil. Industrial Crops and Products 2013;45 208-14.

[49] Sterk B, Leeuwis C, van Ittersum MK. Land use models in complex societal problem solving: Plug and play or networking? Environmental Modelling & Software 2009;24(2) 165-72.

[50] Verburg PH, Schot PP, Dijst MJ, Veldkamp A. Land use change modelling: current practice and research priorities. GeoJournal 2004;61(4) 309-24.

[51] Clark WC, Tomich TP, van Noordwijk M, Guston D, Catacutan D, Dickson NM, et al. Boundary work for sustainable development: Natural resource management at the Consultative Group on International Agricultural Research (CGIAR). Proceedings of the National Academy of Sciences 2011. doi: 10.1073/pnas.0900231108. www.pnas.org/content/early/2011/08/11/0900231108.full.pdf+html (assessed September 26, 2014)

[52] Van Noordwijk M, Lusiana B, Leimona B, Dewi S, Wulandari D, eds. Negotiation-support toolkit for learning landscapes. Volume II. Bogor, Indonesia: World Agroforestry Centre (ICRAF) Southeast Asia Regional Program. http://www.asb.cgiar.org/Publications%202014/ Books/Negotiation%20support%20tool%20kit.pdf (assessed September 26, 2014).

[53] van Noordwijk M. Scaling trade-offs between crop productivity, carbon stocks and biodiversity in shifting cultivation landscape mosaics: the FALLOW model. Ecological Modelling 2002;149(1) 113-26.

[54] van Noordwijk M, Poulsen JG, Ericksen PJ. Quantifying off-site effects of land use change: filters, flows and fallacies. Agriculture, Ecosystems & Environment 2004;104(1) 19-34.

[55] Lippe M, Thai Minh T, Neef A, Hilger T, Hoffmann V, Lam N, et al. Building on qualitative datasets and participatory processes to simulate land use change in a mountain watershed of Northwest Vietnam. Environmental Modelling & Software 2011;26(12) 1454-66.

[56] Lusiana B, van Noordwijk M, Cadisch G. Land sparing or sharing? Exploring livestock fodder options in combination with land use zoning and consequences for livelihoods and net carbon stocks using the FALLOW model. Agriculture, Ecosystems & Environment 2012;159 145-60.

[57] Lusiana B, van Noordwijk M, Suyamto D, Mulia R, Joshi L, Cadisch G. Users' perspectives on validity of a simulation model for natural resource management. International Journal of Agricultural Sustainability 2011;9(2) 364-78.

[58] Dalal-Clayton B, Dent D, Dubois O. Rural Planning in Developing Countries: supporting natural resource management and sustainable livelihoods. London: Earthscan; 2003.

[59] Wang J-J, Jing Y-Y, Zhang C-F, Zhao J-H. Review on multi-criteria decision analysis aid in sustainable energy decision-making. Renewable and Sustainable Energy Reviews 2009;13(9) 2263-78.

[60] Amigun B, Musango J, Brent A. Community perspectives on the introduction of biodiesel production in the Eastern Cape Province of South Africa. Energy 2011;36(5) : 2502-8.

[61] García V. G., Bartolomé M. M. Rural electrification systems based on renewable energy: The social dimensions of an innovative technology. Technology in Society 2010;32 303-11.

[62] Barry M-L, Steyn H, Brent A. Selection of renewable energy technologies for Africa: Eight case studies in Rwanda, Tanzania and Malawi. Renewable Energy 2011;36(11) 2845-52.

[63] Winkler, B. Monitoring and Evaluation of the Integrated Renewable Energy Potential Assessment - A case study in Mgwenyana, rural Eastern Cape, South Africa. MSc Thesis University of Hohenheim. Department for Biobased Products and Energy Crops and the Department for Gender and Nutrition Stuttgart, Germany; 2013.

[64] Chambers R. The origins and practice of participatory rural appraisal. World development 1994;22(7) 953-69.

[65] Hart T. Demystifying participatory research and its role in development. Occasional Paper No. 21. Department of Sociology and Social Anthropology, University of Stellenbosch, South Africa 2008. http://hdl.handle.net/10019.1/45891 (assessed April 7, 2012).

[66] Saaty TL. How to make a decision: the analytic hierarchy process. European journal of operational research 1990;48(1) 9-26.

[67] van Noordwijk M, Lusiana B van Noordwijk M., Lusiana, B. WaNulCAS, a model of water, nutrient and light capture in agroforestry systems. Agroforestry Systems 1999;43 217-42.

[68] Pinto LFG., Bernardes MS., van Noordwijk M., Pereira AR., Lusiana B., Mulia R. Simulation of agroforestry systems with sugarcane in Piracicaba, Brazil. Agricultural Systems 2005;86 275-92.

[69] Pansak W, Hilger T, Lusiana B, Kongkaew T, Marohn C, Cadisch G. Assessing soil conservation strategies for upland cropping in Northeast Thailand with the WaNuLCAS model. Agroforestry Systems 2010;79(2) 123-44.

[70] Walker A, Mutuo P, van Noordwijk M, Albrecht A, Cadisch G. Modelling of planted legume fallows in Western Kenya using WaNuLCAS.(I) Model calibration and validation. Agroforestry Systems 2007;70(3) 197-209.

[71] Leflaive X, Witmer M, Martin-Hurtado R, Bakker M, Kram T, Bouwman L, et al. Water. In: OECD Environmental Outlook to 2050. OECD Publishing; 2012: 207-274, DOI: dx.doi.org/10.1787/9789264122246-en.

[72] IPCC. Climate Change: Synthesis Report 2007 - Contribution of Working Groups I, II and III to the Fourth Assessment Report of the Intergovernmental Panel on Climate Change.2008. www.ipcc.ch/publications_and_data/publications_ipcc_fourth_assessment_report_synthesis_report.html (assesed September 26, 2014).

[73] Kageyama PY, Gandara FB. A Biodiversidade Brasileira e a Questão das Mudanças Globais. Brasília: UNESCO, IBECC; 2008.

[74] Hoekstra A., editor. Virtual water trade: Proceedings of the international expert meeting on virtual water trade. Delft, the Netherlands: IHE; 2003. http://www.waterfootprint.org/Reports/Report12.pdf (assessed September 26, 2014)

[75] Mekonnen M, Hoekstra A. The green, blue and grey water footprint of crops and derived crop products. Hydrology and Earth System Sciences Discussions 2011;8(1): 763-809.

[76] Haberl H, Erb K-H, Krausmann F. Human Appropriation of Net Primary Production: Patterns, Trends, and Planetary Boundaries. Annual Review of Environment and Resources 2014;39(1) 363-91. http://www.annualreviews.org/doi/pdf/10.1146/annurev-environ-121912-094620

[77] Erb K-H, Krausmann F, Lucht W, Haberl H. Embodied HANPP: Mapping the spatial disconnect between global biomass production and consumption. Ecological Economics 2009;69(2) 328-34.

[78] Ellis EC. Sustaining biodiversity and people in the world's anthropogenic biomes. Current Opinion in Environmental Sustainability 2013;5(3) 368-72.

Chapter 6

Challenges and Opportunities in Estimating the Value of Goods and Services in Temperate Grasslands — A Case Study of Prairie Grasslands in Manitoba, Canada

Suren Kulshreshtha, Michael Undi, Jing Zhang, Mohammad Ghorbani, Karin Wittenberg, Ashley Stewart, Esther Salvano, Ermias Kebreab and Kimberly Ominski

Additional information is available at the end of the chapter

http://dx.doi.org/10.5772/59899

1. Introduction

Although the primary function of agricultural lands is to supply commodity outputs such as food, fibre and other raw materials for industrial use, these lands are also a source of many outputs that are not commodities. The non-commodity products are jointly produced and exhibit public goods[1] characteristics. Although more commonly known as externalities of agricultural production, together with the primary production functions are referred to as 'multifunctionality' [1]. Now there is a wide recognition of the services provided by various types of ecosystems [2].

Multifunctionality is important to recognize in the context of human well-being, as it depends on goods and services provided by nature in association with other forms of capital resources. These other capitals include: human capital[2], social capital[3], and manufactured capital[4] (Figure 1). In fact, humanity has always depended on the services provided by the biosphere and its

1 A public good has the characteristic that if supplied to one individual it can be provided to others at no extra cost. Furthermore, no individual can be excluded from the use of that good.

2 Refers to investments made in human resources in order to improve productivity.

3 Social capital is the expected collective or economic benefits derived from the preferential treatment and cooperation between individuals and groups, typically through social networks.

4 Manufactured capital is the infrastructure that is needed to produce various products in the economic system.

ecosystems – natural capital [3]. In order to achieve sustainability, all types of capital are required, although natural capital and economic capital are complementary, and cannot be substituted for each other. Particularly, some of the natural capital cannot even be substituted for any other type of capital (human and manufactured capitals). Furthermore, in some cases, substituting some natural capital may be economically impractical.

An ecosystem service is some attribute of an ecosystem which provides value to humankind [4]. These services are usually related to some attribute of the ecosystem but there is not necessarily a one-to-one correspondence of functions to service. Ecosystems as a provider of goods and services have a value to humankind. These values measure the importance of ecosystem services to members of the society.

The Millennium Ecosystem Assessment [3] has suggested four major types of ecosystem services: provisioning services, cultural services, regulating services, and supporting services that are related to well-being of humans (Figure 2). Provisioning services are the products that the people obtain from the ecosystem. Goods such as food, fibre, water, genetic resources, and others are typical examples. Regulating services include regulation of air quality, climate, water quality, pollination, biological control of diseases, among others. These services are related to productivity of human and non-human systems, and thus have a value. Cultural services result in non-monetary benefits to humans through recreation, aesthetics, and related services. Supporting services are those that are needed for the generation of all the above three types of ecosystem services. Their impact on people is not direct but indirect through these three services.

Figure 1. Types of capital resources relevant to human well-being

Common awareness of concerning trends in climate change and other negative externalities of human activities has recently attracted some attention towards the importance of ecosystem goods and services. Communities and governments have begun to recognize the services that are offered by nature [3]. In response to this need of societies, many researchers and research foundations have started assessing and valuating ecosystems goods and services. Reviewing such studies may shed a light on our way to recognizing the essential goods and services of grasslands in Canada and provide a better understanding of values of these goods and services for society.

http://dx.doi.org/10.5772/59899

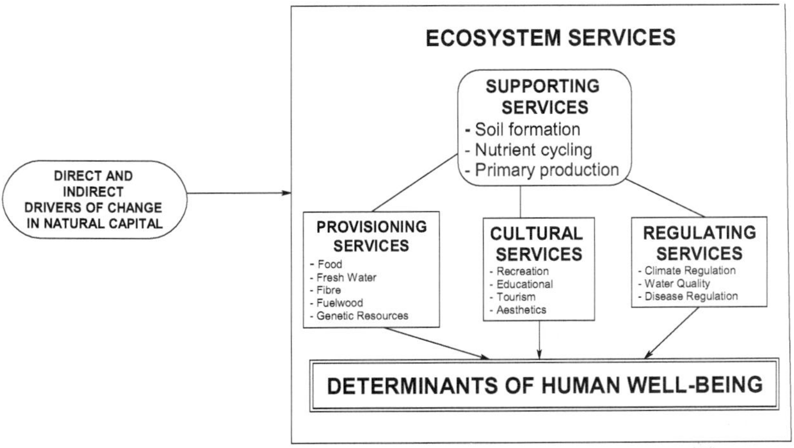

Figure 2. Ecosystems services and ecological goods and services from natural capital (Adapted from [2])

Grasslands are an important part of the rural landscape and are defined as semi-arid areas dominated by herbaceous and shrub vegetation [5]. Worldwide, grasslands fall into three categories, namely, prairie, steppe, and savanna, and cover approximately 3,500 million hectares [6]. On the Canadian prairies, grasslands cover approximately 11 million hectares, extending across southern parts of Alberta, Saskatchewan, and Manitoba, and encompass five prairie eco-regions; Dry Mixed Grass, Mixed Grass, Foothills Fescue, Parkland Northern Fescue, and Tall Grass prairie [7]. Approximately 90% of the Canadian prairie grassland area is grazed by domestic livestock and wildlife [7].

Grasslands have long played a multifunctional role, as a source of feed for wild and domestic herbivores [8,9], and have provided a range of other goods and services (non-commodity goods) such as: management of water resources, carbon sequestration, nitrogen fixation, and recreation [8]. Grasslands are a sustainable source of producing high quality meat and milk [10] and are expected to play an increasingly important role in energy production. The use of perennial grass species, such as switchgrass for ethanol production, could reduce the use of fossil fuels (a positive move in the context of climate change) while also providing producers with an avenue for diversification [11]. Grasslands also play an important role in carbon sequestration. The substantial stocks of carbon sequestered in temperate grassland ecosystems are located largely below ground in the extensive root system of grasses [12, 13]. Perennial forages improve soil quality by reducing erosion and increasing nutrient content of the soil [14]. Grasslands breakdown plant litter and animal wastes and purify water, thus ensuring land and water sustainability and health for future generations [15]. The high plant species richness makes grasslands an ideal habitat for diverse animal populations [8]. Perennial grass cover is important for recreational activities, such as hunting and wildlife viewing [14]. Grasslands provide areas for species reproduction and refuge. For example, ducks need 40% of the landscape as grass in order to achieve nesting success [16]. Grassland flora and fauna are an important genetic resource and provide material for animal and plant breeding and biotechnology [13]. Grasslands are also an important source of biochemical substances that have important medicinal uses [17]. Animal products, such as skins and shells, and flowers from grasslands are important ornamental resources [18]. Thus, the multifunctional role of grasslands provides an important argument for the protection of grasslands as either managed or natural ecosystems.

Ecosystem services through generating ecological goods and services have a value to society, as they contribute to human welfare, both directly and indirectly, and to economic viability through the sustainability of healthy ecosystems [19]. Most goods and services provided by grasslands are not paid for directly and are often overlooked in land-use decision-making processes, resulting in either over-exploitation or inefficient use of grasslands [9]. Establishing an economic value for grassland goods and services increases perception of the importance of grasslands and can lead to improvement in land use in terms of improved grassland management, conservation and protection [9]. Valuation of grasslands can also form the basis for grassland damage assessment and compensation systems [20]. One report on temperate grasslands [20] notes that, although the role of goods and services from temperate grasslands has long been identified as important, the quantitative valuation of such Goods services has not received much attention. A similar sentiment has been expressed through a survey of producers (farmers and ranchers), where recognition of grassland ecosystem services by producers was found to be low – only 25% had awareness of the term 'ecological goods and services' while another 22% indicated some familiarity [21]. To fill this void in past research, this study was undertaken to (a) provide a strategy for assessing the economic value of goods and services from grasslands, (b) identify variables that influence the value of grassland goods and services, and (c) identify gaps in knowledge which require more information to improve the valuation process.

2. Methods

This study was conducted in the Province of Manitoba in Canada. Manitoba, Alberta and Saskatchewan make up the Canadian Prairie Provinces. The Canadian prairies stretch from south-eastern Manitoba to northwestern Alberta [6].

2.1. Concept of value of good/service

Since ecosystem services are a combination of market-based commodities (food and fiber) and non-commodity based goods, their valuation needs to be comprehensive to capture all of these values. Two types of economic valuation are most commonly used: market price method, and non-market valuation. These methods are based on three types of approaches: (1) Revealed willingness to pay; (2) Imputed willingness to pay; and (3) Expressed willingness to pay [22].

Market-based valuation is an example of revealed willingness to pay. People's willingness to pay results in a demand function for that good/service. Here consumers have revealed a preference captured by the curve *DD'* for a given ecosystem service or ecological good or

http://dx.doi.org/10.5772/59899

service (Figure 3). If that good or service is sold through a market, there would be a price established through interaction between buyers and sellers, shown by *PP'*. Similarly the supplier of that good/service would incur certain expenses and willing to offer that good/service only if it covers its cost of producing it. Adding all the sellers' offers for that good/service results in a supply curve for it marked as *SS'*. The area *DP'P* is the benefit to the consumers, called consumer surplus and can be used as the value of that good/service through the use of that ecological good/service. In contrast, the area *PP'S* is the value accruing to producers of that good/service (shaded are in Figure 3), and is equated to be the value of that good/service to them. In the context of grassland, the commercial products are not consumed directly by people, the only relevant value is that to producers.

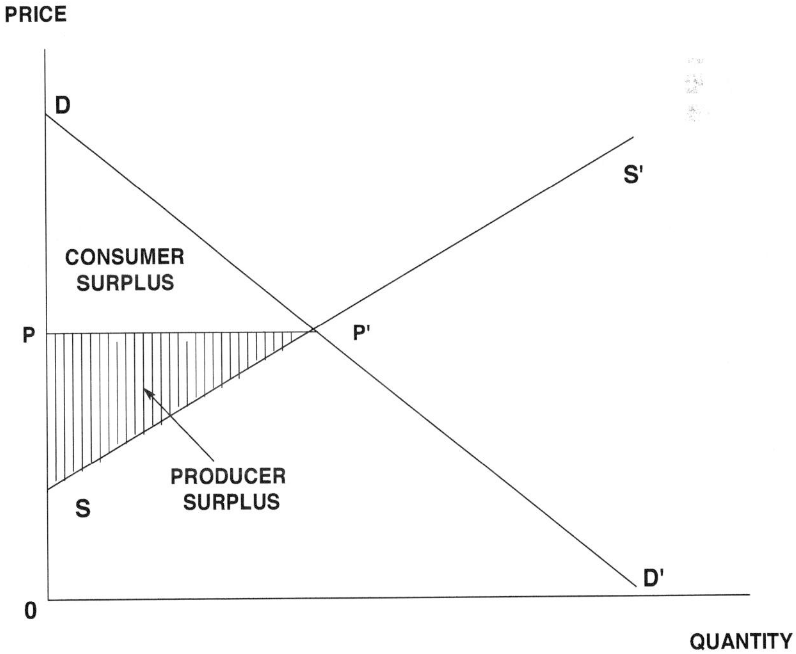

Figure 3. Concept of Consumer and Producers Value of a good/service

Unfortunately many ecological goods and services are not traded through market place. However, for valuation purposes, it is not necessary that an ecosystem service be bought and sold in a market in order to measure its monetary value. What is required under these circumstances is a measure of how much of their purchasing power (dollars) people are willing to give up to enjoy that ecological good/service. Using their revealed willingness to pay, one can use techniques such as Hedonic Pricing method, Travel Cost method or Productivity method to establish a value of the good/service in question. Here the value of a good/service is determined indirectly from the data generated by the marketplace capturing actual market based transactions.

If market based transactions are not available, ecosystem valuation can be based on two other types of approaches: Imputed willingness to pay, and Expressed willingness to pay. These approaches are typically classified as non-market valuation. In the first approach value of an ecosystem service can be imputed from the actions people are willing to take to avoid the adverse effects that could be experienced if that service was lost. Damager Cost Avoided, Replacement cost, and substitute cost are common methods included in this category of non-market valuation. The expressed willingness to pay is through asking people directly what they are willing to pay for an ecosystem service based on a hypothetical scenario. Contingent valuation and Contingent choice methods are included in this category. To undertake non-market valuation, data need to be collected through primary surveys which tends to be costly.

2.2. Process of valuation of Manitoba grassland ecosystem services

The valuation process of grassland goods and services involved several steps. Firstly, detailed information on Manitoba grasslands was collected based on grass type (native, naturalized or tame/seeded), land use (hay, pasture and other) and ownership (private, crown and non-governmental organizations). Tame/seeded grassland was defined as those grasslands which have been cultivated within the past eight years, and are frequently used as part of the crop rotation [23]. Naturalized grasslands are areas that were under cultivation or were seeded to forage and subsequently reverted to grassland, approximately eight to 15 years since last cultivation [23]. Native grasslands were defined as areas that have never been broken, or have been re-established as grassland for such a length of time that native conditions have been restored, more than 15 years since last cultivation [23]. Collection of detailed information on grasslands was followed by identification and valuation of goods and services that could be expected from Manitoba grasslands, as listed in Table 1.

Although a longer list of ecosystem services has been proposed in literature, 21 ecosystem services were identified as being relevant to grassland ecosystems. Of these ten services were excluded from estimation for reasons related to either lack of importance or non-availability of information. The remaining eleven services included all four ecosystem functions. Provisioning services included forage production from grassland, a commercial product for which markets do exist. Under regulating services, six services were identified, including carbon sequestration (thereby affecting gas regulation function), nutrient cycling, water regulation, soil erosion control, soil formation, and water treatment. Recreation and aesthetics was identified as the major cultural service of the grassland ecosystem, whereas refugium services were identified under the supporting services category.

http://dx.doi.org/10.5772/59899

Ecosystem Function	Detailed goods and services	Relevant to grassland in Manitoba	Valuation method
Provisioning services	Forage production	Yes	Market price
	Genetic resources	Nil to ignorable	-
	Medicinal resources	No data	-
	Raw materials	No data	-
	Ornamental resources	No data	-
Regulation services	Carbon sequestration (Gas Regulation)	Yes	Market price –based on carbon trading
	Climate regulation	Nil to ignorable (maybe some local effect, no data)	
	Disturbance prevention	Nil to ignorable	-
	Water regulation	Yes	Value transfer
	Water filtration/water supply	Nil to ignorable	-
	Soil retention/erosion control	Yes	Value transfer
	Soil formation	Yes	Value transfer
	Nutrient cycling	Yes	Market price of accumulated Nitrogen
	Waste treatment	Yes	Value transfer
	Pollination	Yes	30% of market price of food production of grassland relies on pollination
	Biological control	Yes	Value transfer
Supporting services	Wildlife habitat (refugium function)	Yes	Value transfer
	Nursery function	Nil to ignorable	-
Cultural services	Recreation and aesthetics	Yes	Value transfer
	Cultural and artistic information	No data	-
	Spiritual and historic information	No data	-

Table 1. Goods and services from natural ecosystems and methodology of estimation adopted.

Two methods were used to value identified goods and services from the Manitoba grassland ecosystem. A market-based approach was used for goods and services that are traded in the open market. For goods and services that are not traded on the open market, studies that have attempted to value ecosystem goods and services in similar eco-zones were reviewed, and using benefit transfer, values obtained from these studies were applied to non-market goods and services in Manitoba grasslands. Benefit transfer, which in other cases is called environmental value transfer, is related to the process by which a value or demand function of a characteristic or a set of environmental characteristics obtained from each valuation method in a location (original location) can be used to evaluate environmental values in another location (location transfer).

Using the estimates obtained from previous studies to evaluate the costs (or benefits) of new projects, environmental laws or other policies, is common to cost-benefit analysis and public decision making. Benefit transfer approaches are generally recommended and applied by the various institutes for economic valuation of environmental effects. Moreover, because of resource constraints and cost effectiveness, benefit transfer is recommended [24]. In fact, analysts can rarely provide the conditions and facilities of original studies. Therefore, when performing a complete study, transfer studies may provide an economical method to guidance of a researcher [25].

Sometimes the benefit transfer approach is not essentially considered as a methodology, but simply considered as transfer of estimates from one location to other location [25]. Some authors [26] believe that transfer studies involve all advanced skills required to the main research. Therefore, transfer analysts should have high judgment and innovation power of manipulating the existing data and provide results to decision-makers. They should also clearly show the relative roles of data and assumptions and help decision makers to understand the intrinsic uncertainty resources of estimates. Despite the widespread use of this approach, few professional studies exist on how the transfer of data and estimates should be done for grassland ecosystems.

In this study, the benefit transfer approach was used but was also subjected to a sensitivity analysis to account for market price fluctuations or uncertainties in benefit transfer values from other studies. The total economic value of Manitoba grasslands was obtained by summing economic values of market and non-market goods and services.

2.3. Market-based valuation

Under market-based valuation, a link between the environmental (ecosystem) service (and ecological goods and services generated by it) and society's preference is developed. If the good is commercial in nature, it is bought and sold through the marketplace. Its demand reflects social preference (or value). If market price for a certain grassland service in the marketplace exists, the price is directly used to evaluate the goods and services [27]. Market-based valuation was used to assess the value of perennial forage production, carbon sequestration, and nutrient cycling.

2.3.1. Perennial forage production

Data on grassland area, forage yield and forage prices were obtained from consultations with individuals with in-depth knowledge of Manitoba grasslands (Bill Gardiner, MAFRI; Glenn

Friesen, MAFRI; Rick Andrews, Ducks Unlimited Canada; Wybo Vanderschuit, Riding Mountain National Park) and other sources [23]. The yield of native and naturalized hay was estimated to be 3.92 t/ha/yr whereas that for tame/seeded hay was estimated to be 5.91 t/ha/yr (Glenn Friesen, Manitoba Provincial Forage Specialist, personal communication). The tame/seeded hay yield is the average of the alfalfa and alfalfa-grass yields. It was assumed that grass hay yields were comparable to alfalfa/grass hay yields. The yield for forage seed production (0.38 t/ha/yr) was the average of the yield in 2005 and 2006 [24]. The price (2004-2008 average) utilized for native and naturalized hay was $0.042/kg (Glenn Friesen, personal communication). Grass and alfalfa/grass hay grown on tame/seeded grasslands was valued at $0.055/kg and $0.075/kg, respectively. Forage seed was valued at $1.10/kg (average price for 2005-2006). Native and naturalized pasture was valued at $34.37 ha/yr and tame/seeded pasture was valued at $101.33 ha/yr (Glenn Friesen, personal communication). Energy production, an important direct value of forages, was not included in the valuation of Manitoba grasslands as there are currently no facilities for biofuel production from grass in Manitoba. In fact, perennial grasses that can be used to produce cellulosic ethanol are not yet grown on a commercial scale [11].

2.3.2. *Carbon sequestration*

The economic value of carbon sequestration was based on sequestration estimates from the Pilot Analysis of Global Ecosystems (PAGE) [5]. The PAGE carbon sequestration estimate for grasslands ranges from 100 to 300 t/ha/yr, with mid-latitude grasslands having lower carbon sequestration than high- and low-latitude grasslands [9]. This study adopted a carbon sequestration value of 105 t/ha/yr for valuation of Manitoba grasslands, as suggested to be appropriate for Canadian grasslands [28]. Due to lack of data on rate of carbon sequestration for various types of grasslands, the same level was assumed for all grasslands in Manitoba. Total amount of carbon sequestered by Manitoba grasslands was estimated at 250.5 million t/yr (Table 2). About 64% of this amount is sequestered by tame or seeded pastures, and another 35% by native pastures in the province.

	Grassland type			
	Native	Naturalized	Tame/seeded	All
Carbon sequestration[1], t/ha/yr	105	105	105	
Area (c), ha	826,334	19,926	1,539,400	2,385,660
Total carbon sequestration, t/yr	86,765,070	2,092,230	161,637,000	250,494,300

[1]Suggested value for Canadian grasslands [32].

Table 2. Calculation of level of carbon sequestration.

Valuation of the carbon sequestered by grasslands is not a simple matter since it is not traded in a fully functional market place. A close approximation to a market is the Chicago Climate

Exchange where in 2009 carbon was trading at US$2.10 per ton. Converting it using current exchange rates leads a value of $2.67 per ton in Canadian funds. This resulted in a total value of $668.8 million. However, during the past, exchange rates between US and Canadian dollar have fluctuated. In addition trading value of carbon at the Chicago Climate Exchange has also fluctuated from $1.60 to $2.15 per ton in US dollar. To see the change in the value of carbon sequestration, a sensitivity analysis was undertaken. The value of carbon sequestration ranged from $508 to $683 million per year (Table 3). On account of higher level of carbon sequestration, value of non-native grassland was higher ranging from $326 to $438/ha/yr compared to native grassland. If one argues that rate of sequestration or its unit value as shown in these tables can vary, further sensitivity analysis needs to be undertaken, which is presented in a later section.

	Exchange rate	Trading Price		Total value	Value (Can $/ha/yr)	
		2009 US $/ton	2009 Can $/ton	2009 Can $	All grasslands	Excluding native grasslands
Base scenario	1.2718	2.10	2.67	668,819,781	280.35	428.91
Lower C price scenario	1.2718	1.60[1]	2.03	508,503,429	213.15	326.10
Higher C price scenario	1.2718	2.15[2]	2.73	683,849,439	286.65	438.55
Strong Canadian dollar	0.9984[2]	2.10	2.10	526,038,030	220.50	337.35

[1]Lowest trading price between 2003-2010 at Chicago Climate Exchange.

[2]The exchange rate on April 21, 2010.

Table 3. Sensitivity analysis of total value of carbon sequestration in Manitoba grasslands.

2.3.3. *Nutrient cycling*

The value of increasing soil nitrogen was determined as a product of estimated amount of nitrogen accumulation in the soil, area of grassland and the value of the accumulated nitrogen. Utilizing a 10:1 ratio of carbon to nitrogen accumulation [29] resulted in accumulation of 0 t nitrogen ha/yr in native grassland, and 0.047 and 0.056 t nitrogen ha/yr in naturalized and tame/seeded grassland, respectively. The accumulated nitrogen was valued at $1.32/kg, the value of urea fertilizer. This resulted in a value for nutrient cycling of $81.47/ha.

2.4. Non-market-based valuation

Most goods and services in grasslands are not traded in the marketplace and require valuation techniques that reflect their existence outside the market system [27]. Techniques for the valuation of such non-market goods and services have been discussed above. Most of these techniques are time-consuming and require considerable resources to complete. Selection of benefit transfer as the method of choice in this study was based on cost-effectiveness and previous application in the valuation of ecosystem goods and services. Benefit transfer is widely applied in the economic valuation of non-market services, often as part of cost-benefit analysis of a new project that has environmental impacts. A legitimate use of benefit transfer should meet the following conditions: a) population of both regions should be similar, b) goods and services in all locations should have about the same characteristics, and, c) initial estimated values should be current because preferences change over time [30].

The benefit transfer values utilized in this study were obtained mainly from two studies [14, 24] and are listed in Table 4. The first study [14] used four case studies from different agricultural regions across Canada to assess the ecological goods and services provided by the natural capital within settled areas. The case studies estimated that the net value of conserving or restoring natural areas varied from $65/ha/yr in the Upper Assiniboine River Basin in eastern Saskatchewan and western Manitoba, $142/ha/yr in the Mill River Watershed in Prince Edward Island, to $195/ha/yr in the Grand River Watershed of Ontario. The transfer values from this study [14] were, in turn, obtained from a report [31]. The second study [24] undertook an evaluation of the economic value of New Jersey's wetlands, marine ecosystems, forests, urban green space, beaches, agricultural land, open fresh water and riparian buffers. The transfer values from this study were only used in situations where no Canadian values exist.

Other important sources of benefit transfer values relevant to the Canadian prairies [32-34] were consulted and appropriate values were selected for this study. Benefit transfer was used to value water regulation, waste treatment, soil erosion control, soil formation, recreation, and wildlife habitat (refugium). The value of water regulation in Manitoba grasslands was calculated from total grassland area (2,385,660 ha) using the benefit transfer value of $5.14/ha/yr (Table 4). Benefit transfer values for soil formation ($10.70/ha/yr) and erosion control ($13.34/ha/yr) were transferred from two studies [24, 35] (Table 4). Waste treatment was estimated utilizing a benefit transfer value of $64.52/ha/yr (Table 4). The value of cultural services was transferred from reference [14]. The value of refugium was transferred from a study [34] which estimated willingness to pay for prairie grassland conservation for burrow owl (an endangered species) in Southern Alberta at approximately $34.07/rural household (Table 4).

In determining the preferred value to be transferred to this study, values derived from a similar eco-zone of grassland in Manitoba were used. In general, values transferred from North America grassland were preferred to the global grassland. Similarly, Canadian grassland values were preferred over the North America grassland. If a choice was available, values from Western Canada were preferred over those from Eastern Canada.

3. Results

Grasslands of various types occupy approximately 2.4 million ha of Manitoba lands (Table 5). Of this area, 64.5% is tame/seeded grassland and 34.6% is native grasslands. Naturalized grasslands make up less than 1% of the total grassland area. Almost half of Manitoba grasslands (54%) are utilized for pasture, while another 44% are utilized for hay production. Most of the grassland area in Manitoba (90%) is privately-owned (Table 5).

Single value $/ha/yr	Value range $/ha/yr	Notes	Source
		Water regulation	
5.14		Improved water quality-decreased sediment in Upper Assiniboine River Basin	[14]
7.14	1.48 – 7.14	Average of grassland in the world	[24]
		Erosion control	
2.97		Estimate is based on reduced wind erosion in Upper Assiniboine River Basin	[14]
13.34	2.97-53.45	Based on prairie soil in Canada	[34]
53.45		Average of grassland in the world	[24]
		Soil formation	
10.7	10.7	Average of grassland in the world	[19]
		Waste treatment	
64.52		Waste treatment services by forests (i.e. removal of phosphorus and nitrogen	[14]
157.03	64.52-157.03	Average of world grassland	[24]
		Recreation & aesthetics	
17.05		Assuming that cropland provides no habitat to game species of wildlife, the average hunting value for lands in permanent cover is $11.91/ha/yr, an upper bound is $23.72/ha/yr. Revenue related to wildlife viewing is about $5.14/ha/yr.	[14]
0.1	0.10-17.05	Based on discrete travel cost study in Alberta	[32]
		Refugium function	
4.6		Individual households are willing to pay $34.07 to conserve grassland habitat for burrowing owl, and there are 321,750 rural households in Manitoba	[33]
0.25	0.20-4.60	$1.48-2.69/household	[14]

[1]Preferred transfer values were derived from a similar eco-zone as grassland in Manitoba. North America grassland values were preferred to global grassland values.

Table 4. Summary of grassland goods and services values[1] reported in other studies and utilized in benefit transfer.

3.1. Market-based valuation

3.1.1. Perennial forage production

Most perennial forage production in Manitoba occurs on native (826,334 ha) or tame/seeded grasslands (1,484,999 ha), with only small amounts of naturalized grassland (18,211 ha) being utilized for this purpose (Table 5). Forages in Manitoba are primarily utilized for grazing and preserved forage (hay and silage) but may also be used for forage seed production (Table 5). The total value of forage production in Manitoba was approximately $524 million/yr (Table 6). Sensitivity analysis of the volatility of forage prices indicated that a 20% increase in the price of hay would increase the value of seeded forage or pasture to $629.5 million/yr. An equivalent decrease in the price of hay would reduce the value of seeded forage/pastures to $419.7 million/yr.

Grassland Type	Ownership	Grassland use (ha)			Sub-total
		Hay	Pasture	Other uses[2]	
Native Grassland[3]	Private	82,537	577,754	137[6]	660,428
	Crown	10,114	152,777	2,999	165,889
	NGO	-	17	-	17
	Sub-total				**826,334**
Naturalized Grassland[4]	Private	-	-	137[6]	137
	Crown	4,098	14,113	-	18,211
	NGO	-	-	1,578	1,578
	Sub-total				**19,926**
Tame/seeded Grassland[5]	Private	945,308	498,312	53,551	1,497,171
	Crown	5,008	36,371	-	41,379
	NGO	-	-	850[6]	850
	Sub-total				**1,539,400**
Grand total					**2,385,660**

[1]Compiled from several sources [24,35-37] and personal communication with Bill Gardiner (MAFRI), Glenn Friesen (MAFRI), Rick Andrews (Ducks Unlimited Canada), and Wybo Vanderschuit (Riding Mountain National Park).

[2]Mainly forage seed production, green space and aesthetic appeal.

[3]Grasslands which have been cultivated within the past eight years and are frequently used as part of the crop rotation.

[4]Areas that were under cultivation or were seeded to forage and subsequently reverted to grassland (approximately eight to 15 years since last cultivation).

[5]Areas that have never been broken, or have been re-established as grassland for such a length of time that native conditions have been restored (> 15 yr since last cultivation).

[6]Conservation purposes (Rick Andrews, Ducks Unlimited, Canada, personal communication).

Table 5. Area of Manitoba grasslands by grass type, use and ownership[1].

	Production			
Grassland type[2]	Hay	Pasture	Forage seed	Total
Native	$16,767,375	$25,109,664	0	
Naturalized	$88,984	$485,111	0	
Tame/Seeded	$402,814,401	$54,181,756	$25,166,927	
Total	$419,670,760	$79,776,531	$25,166,927	$524,614,218

[1]Areas of different grassland types are shown in Table 5.

[2]Defined in Table 5.

Table 6. Total annual value[1] of forage and seed production from Manitoba grasslands.

3.1.2. *Carbon sequestration*

With 19,926 ha of naturalized grassland and 1,539,400 ha of tame/seeded grassland in Manitoba, carbon sequestration in Manitoba grassland was estimated at approximately 250.5 million tons annually (Table 5). The average value of carbon sequestration was $280.35/ha/yr for all types of grasses (Table 7). Alternative values that were estimated to account for fluctuations in Canadian-US dollar exchange showed that the total value of carbon sequestration could range from approximately $508.5 million/yr to $683.8 million/yr (Table 7). The average value of carbon sequestration for all grasslands was approximately $213.35 to $286.65/ha/yr or, if native grassland is excluded, $326 to $439/ha/yr (Table 7). The value of carbon sequestration in this study lies between estimates of $267/ha/yr and $469/ha/yr reported in other studies [28,38].

Source	Price ($/ha)	Area (ha)	Total value (2009 Can $)
This study	280.35	2,385,660	668,819,781
[38]	266.94	2,385,660	636,828,080
[28]	468.84	2,385,660	1,118,492,834

Table 7. Sensitivity analysis of the total value of carbon sequestration.

3.1.3. *Nutrient cycling*

Perennial forages can improve land productivity by increasing the nutrient content of soil. For example, inclusion of legumes in pastures will increase soil nitrogen due fixed atmospheric nitrogen being added to the soil [39]. The increase could also be due to the ability of forages to access nitrate from soil depths of more than one meter below the surface. With a total of 1,559,326 ha of naturalized and tame/seeded grassland, the total value of nutrient cycling was estimated at $127.04 million/yr. If the price of nitrogen fluctuates by 20%, the total value of nitrogen will vary between $101.63 and $152.45 million/yr (Table 8).

	Best value[1]		Low estimate		High estimate	
Goods and services	Total ($ million)	$/ha	Total ($ million)	$/ha	Total ($ million)	$/ha
Provisioning services						
Forage production	524.61	219.89	419.69	175.93	629.54	263.87
Regulating services						
Carbon sequestration						
Low CO_2 price	508.50	213.15	508.50	213.15	508.50	213.15
Middle CO_2 price	668.82	280.35	668.82	280.35	668.82	280.35
High CO_2 price	683.85	286.65	683.85	286.65	683.85	286.65
Nutrient cycling	127.04	81.47	101.63	65.18	152.45	97.78
Water regulation	12.26	5.14	3.54	1.48	17.04	7.14
Soil erosion control	31.85	13.34	7.07	2.97	127.51	53.45
Soil formation	25.52	10.70	8.94	3.76	29.60	12.40
Waste treatment	153.92	64.52	153.92	64.52	374.62	157.03
Cultural service						
Recreation and aesthetics	40.67	17.05	0.01	0.10	40.67	4.74
Supporting services						
Refugium function	10.96	4.60	0.25	0.20	0.46	0.37
Low CO_2 price	1,435.33	629.86	1,203.55	527.29	1,880.39	809.93
Middle CO_2 price	1,595.65	697.06	1,363.87	594.49	2,040.71	877.13
High CO_2 price	1,610.68	703.36	1,378.90	600.79	2,055.74	883.43

[1]Estimates are based on 2,385,660 ha of grassland except nutrient cycling (1,559,325 ha).

Table 8. The annual value of goods and services from Manitoba grasslands.

3.2. Non-market-based valuation

The value of water regulation was estimated to be $12.26 million/yr or $5.14/ha/yr (Table 8). The total value for erosion control in Manitoba grasslands was assessed to be $32 million/yr or $13/ha/yr while soil formation by grasslands was estimated to be $26 million/yr (Table 8). Based on 2,385,660 ha of grassland, waste treatment in Manitoba grasslands was valued at approximately $153.92 million/yr (Table 8). Cultural services from grasslands include recreation, aesthetics, and cultural information. Perennial forage cover increases recreational activities such as hunting and wildlife viewing [14]. In riparian areas, perennial forage cover can increase the use of an area for fishing, camping, swimming, and canoeing [14]. Based on approximately 2.4 million ha of grassland and benefit transfer value of $17.05/ha/yr (Table 4), the value of cultural services from Manitoba grasslands was estimated to be $40.67 million/yr

(Table 8). With 321,750 rural households in Manitoba in 2006, the total value for conserving the burrowing owl in Manitoba grasslands was estimated to be $10.96 million/yr or approximately $4.60/ha/yr (Table 4). Using these estimates, the total willingness to pay by Manitoba rural households was estimated to range from $0.25 to $0.46 million/yr or $0.20 to 0.37/ha/yr (Table 8).

3.3. Total value of Manitoba grasslands

The total economic value of Manitoba grasslands varies between $1,204 million/yr ($527/ha/yr) to $2,056 million/yr ($883/ha/yr; Table 8). This range is relatively narrow as a consequence of the sensitivity analysis conducted. The scope of research for values that were derived from benefit transfer was limited to those studies conducted in a similar eco-zone to Manitoba grasslands. Further, a 20% price fluctuation in prices was assumed for values that are derived using market price. The economic value of Manitoba grasslands obtained in this study should be taken as a minimum value which is expected to change as information specific to Manitoba grasslands becomes available.

Among all the estimated values of various ecosystem services, besides the commercial (market-based) values, carbon sequestration is an important ecosystem service (Figure 4). About a third of the total value of Manitoba grasslands is through production of forages and related goods. Under most expected conditions, the ecosystem service most important for this ecosystem is carbon sequestration. At this time, 42% of total value is credited to this service. The third most important ecosystem service for the Manitoba grassland is from waste treatment.

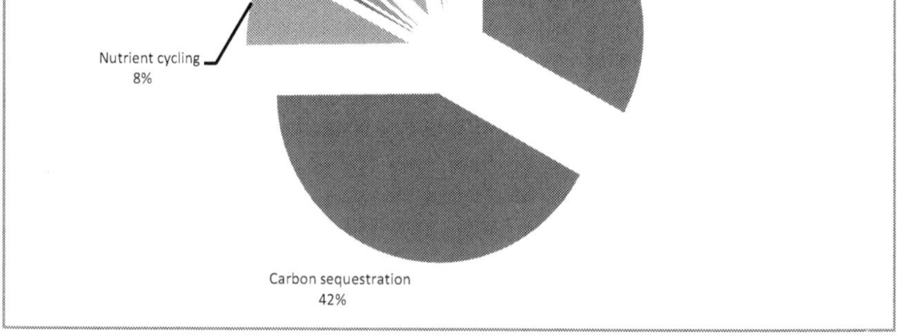

Figure 4. Distribution of total economic value of Manitoba grassland by type of ecosystem service

4. Discussion

This study set out to provide a strategy for estimating the economic value of goods and services from Manitoba grasslands by utilizing market and non-market based approaches. Assigning value to goods and services such as perennial forage production and carbon sequestration that are traded in the open market is a matter of identifying prevailing market values. The market value of goods and services such as forage production will depend on the quality of the goods and services. The value of pasture, for example, should take into account season, pasture plant species, and pasture management since such factors will influence pasture quality. For hay, quality characteristics including nutrient content (protein and energy) as well as organoleptic characteristics, such as color, mold, and dust, to assign value would lead to a more objective price determination. Such a pricing system would give a range in prices for forage production. The strategy of assigning value based on forage quality was not employed in the current study due to inadequate information. Carbon sequestration is influenced by location of the grasslands [28] and by management practices that are imposed on the grasslands [10]. Carbon sequestration estimates reported in these studies were not measured directly and the differences in estimates among studies suggest a need for direct carbon sequestration measurements in grasslands.

Benefit transfer was used to valuate non-market goods and services in Manitoba grasslands. Challenges in applying benefit transfer analysis are mainly related to difficulties in identifying and selecting suitable studies for comparison, in combining data and in transferring data [40, 41]. It is also important to note that most studies are not designed with the aim of transferring [40, 41]. A more complex approach, which we did not use in this study, is to use meta-analysis to systematically analyze the impact of a study on estimated values [26]. The use of the benefit transfer method is not universally accepted and has been questioned. In most cases, the original studies will have valued small changes in specific and localized components of individual ecosystems, which makes it incorrect to extrapolate value estimates obtained from these localized scenarios to a much larger scale [26]. Benefit transfer has also been criticized as being dirty, quick and ugly [42]. Some economists do not consider benefit transfer as a methodology, but simply consider it as transfer of estimates from one location to other location [25]. Others (43) consider the transfer of valuations from one ecological and social context to another as dangerous because ecosystem values are highly dependent on location. Until more appropriate methods are utilized to value ecosystem goods and services, benefit transfer will likely remain the method of choice.

Genetic, medicinal, and ornamental resources, water supply and cultural services, such as cultural and artistic information, and spiritual and historic information were not included in the valuation of Manitoba grasslands due to lack of information. Such goods and services tend to be site-specific and values obtained within the region of interest, in this case, prairie grasslands, would be more appropriate. Since no valuation studies for these goods and services were identified, primary data collection would be the only method to collect such information. Techniques that can be utilized to value these goods and services have been suggested [27]. A market-based valuation (direct market pricing) is appropriate to value genetic, medicinal, and ornamental resources, water supply, and nursery function although other techniques such as factor income, replacement cost, and contingent valuation can also be applied [27]. For cultural services, techniques such as contingent valuation, travel cost, and hedonic pricing are important [27]. With growing emphasis on valuation of grassland goods and services, there is a need to conduct specific studies that will provide values for such goods and services. Such information will further improve the value of grasslands, thus further emphasizing the importance of maintaining productive grasslands.

5. Conclusion

Grasslands have a multi-functional role, providing food for herbivores as well as other goods and services such as carbon sequestration, nature conservation, and recreation. The goods and services provided by grasslands have direct and indirect monetary value. This study was conducted to a) provide a strategy for assessing the economic value of goods and services from grasslands, b) identify variables that influence the value of grassland goods and services, and c) identify gaps in knowledge which require more information to improve the valuation process. The study was conducted in three stages. Firstly, information regarding grasslands in the Province of Manitoba, Canada was collected based on grass type (native, naturalized or tame/seeded), land use (hay, pasture and other) and ownership (private, crown and non-governmental organizations). This was followed by identification of goods and services that could be expected from these grasslands. The identified goods and services were then valuated. Market prices were utilized to value grassland goods and services where transactions occur in the marketplace while the benefit transfer method was used to infer monetary values of those goods and services that are not typically sold through the marketplace. Sensitivity analysis was performed to account for market price fluctuation or uncertainties in benefit transfer. The total economic value of Manitoba grasslands was derived by summing economic values of goods and services for which supporting data was available.

Although the study identified 21 goods and services provided by Manitoba grasslands, only perennial forage production, carbon sequestration, nutrient cycling, water regulation, waste treatment, soil erosion control, soil formation, recreation, and wildlife habitat were included in the valuation. Genetic, medicinal, and ornamental resources, water supply, nursery function, biological value, and cultural services, while important, could not be assigned monetary value due to lack of data. The total economic value of Manitoba grasslands was estimated to be approximately $1,436 million/yr ($630 ha/yr), with a range of $1,203 million/yr ($527 ha/yr) to $1,880 million/yr ($810 ha/yr). The two most highly valued goods and services in Manitoba grasslands were perennial forage production and carbon sequestration. Multidisciplinary research, focusing on economic valuation of non-market goods and services will provide more relevant transfer values than those obtained from other ecosystems and will greatly improve estimates of grassland value.

This study shows that goods and services in complex ecosystems such as temperate grasslands can be valued using market and non-market based valuation methods. Improvements in market-based valuation can be accomplished by local research that will more precisely quantify goods and services such as carbon sequestration and assess forage value based on forage quality. The major challenge in valuation of grasslands lies in the development and use of methods to improve valuation of non-market goods and services. While benefit transfer, as utilized in this study, gives estimates of the value of non-market goods and services, local grassland research will be required to ground truth benefit transfer values. The limited number of studies reporting values of non-market goods and services in North America made use of benefit transfer in the current study somewhat complex. This study could not access information on the value of goods and services such as genetic, medicinal, and ornamental resources, water supply, and cultural services, which will be required for a complete valuation of grasslands. Ultimately, the valuation of grasslands will require economic values of goods and services obtained directly from grasslands. In this respect, investment into multidisciplinary research focusing on the economic valuation of grassland goods and services will provide more relevant transfer values than those obtained from other ecosystems and will greatly improve grassland valuation.

Acknowledgements

This study was funded by Beef Cattle Research Council (BCRC), Manitoba Beef Producers, Canada Manitoba Agri-food Research and Development Initiative (ARDI), MITACS, Green-Cover, and Manitoba Rural Adaptation Council (MRAC). The authors would like to extend sincere appreciation to Bill Gardner (MAFRI), Glen Friesen (MAFRI), Don Flaten (University of Manitoba), Rick Andrews (Ducks Unlimited), and Wybo Vanderschuit (Riding Mountain National Park) for sharing their insights on Manitoba grasslands.

Author details

Suren Kulshreshtha[1*], Michael Undi[2], Jing Zhang[1], Mohammad Ghorbani[1], Karin Wittenberg[2], Ashley Stewart[3], Esther Salvano[4], Ermias Kebreab[5] and Kimberly Ominski[2]

*Address all correspondence to:

1 University of Saskatchewan, Saskatoon, Canada

2 University of Manitoba, Winnipeg, Canada

3 Shur-Gro Farm Services Ltd., Waskada, Manitoba, Canada

4 Manitoba Agriculture, Food and Rural Initiatives, Winnipeg, Canada

5 University of California, Davis, CA, USA

References

[1] Abler D. Multifunctionality in agriculture: evaluating the degree of jointness, policy implications. In Organization for Economic Cooperation. Multifunctionality in Agriculture: Evaluating the degree of Jointness, Policy Implications. Paris. 2008.

[2] Woodward RT, Wui, Y. The economic value of wetlands services: a meta-analysis. 2001;37:257-270.

[3] Alcamo J. [et al.]. Ecosystems and human wellbeing. A report of the conceptual framework working group of the Millennium Ecosystem Assessment. Washington, DC: Island Press. 2003.

[4] Edward-Jones G. Davies B. Hussain S. Ecological economics – an introduction. London: Blackwell Science Ltd. 2000.

[5] White R, Murray S, Rohweder M. Grassland ecosystems. Pilot analysis of global ecosystems. World Resources Institute. Washington, D.C. 2000. Available online at Web site: www.wri.org/publication/pilot-analysis-global-ecosystems-grassland-ecosystems. (accessed June 25, 2010).

[6] Carlier L, Rotar J, Vlahova M, Vidican R. Importance and functions of grasslands. Notulae Botanicae Horti AgrobotaniciCluj-Napoca 2009;37:25-30. Available online at Web site: www.notulaebotanicae.ro (accessed September 1, 2011).

[7] Bailey AW, McCartney D, Schellenberg MP. Management of Canadian Prairie Rangeland. Agriculture and Agri-Food Canada. 2010. Available online at Web site: www.agr.gc.ca/scienceandinnovation (accessed January 8, 2012).

[8] Wiltshire K, Delate K, Wiedenhoeft M. Socio-cultural aspects of cow-calf operation persistence in a peri-urban county in Iowa. Renewable Agriculture and Food Systems 2010;26(1):60-71. doi:10.1017/S1742170510000505 (accessed July 3, 2013).

[9] Boval M, Dixon RM. The importance of grasslands for animal production and other functions: a review on management and methodological progress in the tropics. Animal 2012;6: 748-762. doi: 10.1017/S1751731112000304 (accessed September 17, 2012).

[10] Liu J, Diamon J. China's environment in a globalizing world. Nature 2005;435:1179-1186.

[11] Bourlion N, Janssen L, Miller M. Economic analysis of private and public benefits of corn, switchgrass and mixed grass systems in Eastern South Dakota. Renewable Agriculture and Food Systems 2013. available on CJO2013doi:10.1017/S1742170513000239 (accessed July 3, 2013).

[12] Janzen HH, Campbell CA, Izaurralde RC, Ellert BH, Juma N, McGillWB, Zentner RP. Management effects on soil C storage on the Canadian prairies. Soil and Tillage Research 1998;47:181-195.

[13] Jones MB, Donnelly A. Carbon sequestration in temperate grassland ecosystems and the influence of management, climate and elevated CO_2. New Phytologist 2004;164:423-439.

[14] Olewiler N. The value of natural capital in settled areas of Canada. Ducks Unlimited Canada and the Nature Conservancy of Canada. 36 pp. 2004. Available online at Web site: www.ducks.ca/aboutduc/news/archives/pdf/ncapital.pdf (accessed January 23, 2008).

[15] Tilman D, Cassman KG, Matson PA, Naylor R, Polasky S. Agricultural sustainability and intensive production practices. Nature 2002;418:671-677.

[16] Nelson D. It's the Ecosystem, Stupid. Delta Waterfowl Magazine. 2004. Available online at Web site: www.deltawaterfowl.org/magazine/2004_01/01_ecosystem.php (accessed June 22, 2009).

[17] Hönigová I, Vačkář D, Lorencová E, Melichar J, Götzl M, Sonderegger G, Oušková V, Hošek M, Chobot K. Survey on grassland ecosystem services. Report of the European Topic Centre on Biological Diversity. 2012. Nature Conservation Agency of the Czech Republic. pp 78. Available online at Web site: www.teebweb.org/wp-content/uploads/2013/01/Survey-on-grassland-ES_2011_final-report_ISBN.pdf (accessed August 2, 2013).

[18] Saskatchewan Wetland Conservation Corporation. Native Prairie Stewardship Fact Sheets. 2002. ISBN # 1-896793-04-5. Available online at Web site: www.swa.ca/Publications/Documents/NativePrairieStewardship8FactSheetsHarvestMarketNative-Seeds.pdf. (accessed September 22, 2010).

[19] Costanza R, D'arge R, de Groot R, Farber S, Grasso M, Hannon B, Limburg K, Naeem N, O'neill RV, Paruelo J, Raskin RG, Sutton P, Van Den Belt M. The value of the world's ecosystem services and natural capital. Nature 1997;387:253-260.

[20] Heidenreich B. What are global temperate grasslands worth? A case for their protection. 2009. Prepared for The World Temperate Grasslands Conservation Initiative. Available online at Web site: cmsdata.iucn.org/downloads/grasslandssocioeconomicreport.pdf (accessed February 13, 2012).

[21] Environics. 2006. National survey of farmers and ranchers: ecological goods and services. Toronto.

[22] King, DM, Mazzotta MJ. Ecosystem valuation. Available online at Web site: www.ecosystemvaluation.org/benefit_transfer.htm. (accessed 25 July 2014).

[23] Statistics Canada. Census of Agriculture. Agriculture Overview, Can ada and the Provinces – Table 1.5. Land Use, Census years 2001 and 2006. 2006a. Available online at Web site: www.statcan.gc.ca/pub/95-629-x/1/4123822-eng.htm#46 (accessed January 23, 2008).

[24] Costanza R, Wilson M, Troy A, Voinov A, Liu S, D'Agostino J. The Value of New Jersey's Ecosystem Services And Natural Capital. 2006. Gund Institute for Ecological

Economics and New Jersey Department of Environmental Protection. University of Vermont. Burlington, Vermont.

[25] Garrod G, Willis KG. Economic valuation of the environment. 1998. Cheltenham: Edward Elgar Publishing.

[26] Desvouges, DS, Johnson, FR, Spencer, HS. Environmental analysis with limited information. 1998. Cheltenham: Edward Elgar Publishing.

[27] de Groot RS, Wilson MA, Boumans RMJ. A typology for the classification, description and valuation of ecosystem functions, goods and services. Ecological Economics 2002;41:393-408.

[28] Smith WN, Desjardins RL, Grant B. Estimated changes in soil carbon associated with agricultural practices in Canada. Canadian Journal of Soil Science 2001;81:221-227.

[29] Janzen HH, Beauchemin KA, Bruinsma Y, Campbell CA, Desjardins RL, Ellert BH, Smith EG. The fate of nitrogen in agroecosystems: an illustration using Canadian estimates. Nutrient Cycling in Agroecosystems 2003;67:85-102.

[30] Brookshire DS, Neill HR. Benefit transfers: Conceptual and empirical issues. Water Resources Research 1992;28:651-655.

[31] Belcher K, Edwards CK, Gray B. Analysis of Economic Instruments: Conservation Cover Program. Case Studies: Grand River Watershed, Ontario, Upper Assiniboine River Basin, Saskatchewan and Manitoba, and Mill River Watershed, Prince Edward Isalnd. 2001. Ecological Fiscal Reform and Agricultural Landscapes. Final Report.

[32] Boxall P. The economic value of lottery-rationed recreational hunting. Canadian Journal of Agricultural Economics 1995;43:119-131.

[33] Atakelty H, Adamowicz V, Boxall P. Complements, substitutes, budget constraints and valuation. Environmental and Resource Economics 2000;16:51-68.

[34] Agriculture and Agri-Food Canada. Prairie soils: The case for conservation. 2003. Available online at Web site: www.agr.gc.ca/pfra/pub/prairiesoils_e.htm (accessed January 30, 2009).

[35] Kulshreshtha S, Pearson GG. Estimation of cost recovery levels on federal community pastures under joint private and public benefits. 2002. A report prepared for land management division, land management and diversification service, Prairie Farm Rehabilitation Administration, Regina, SK. Canada.

[36] Manitoba Agriculture, Food and Rural Initiatives. Manitoba agriculture yearbook. 2007. Available online at Web site: www.gov.mb.ca/agriculture/statistics/yearbook2006/2006_manitoba_agriculture_yearbook.pdf (accessed January 23, 2008).

[37] Statistics Canada. Census of Agriculture. Agriculture Overview, Canada and the Provinces – Table 1.5. Hay and field crops. Census years 2001 and 2006. 2006b. Avail-

able online at Web site: www.statcan.gc.ca/pub/95-629-x/1/4123806-eng.htm#46 (accessed January 23, 2008).

[38] Wilson S. Ontario's wealth, Canada's future: Appreciating the value of the Greenbelt's eco-services. 2008. David Suzuki Foundation, Vancouver, Canada. Available online at Web site: www.ecosystemvaluation.org/benefit_transfer.htm (accessed May 27, 2010).

[39] Kelner DJ, Vessey JK, Entz MH. The nitrogen dynamics of 1-, 2- and 3-year stands of alfalfa in a cropping system. Agriculture, Ecosystems and Environment 1997;64:1-10.

[40] Brouwer R. Environmental value transfer: State of the art and future prospects. Ecological Economics 2000;32:137-152.

[41] Bockstael NE, Freeman AM, Kopp RJ, Portney PR, Smith VK. On measuring economic values for nature. Environmental Science and Technology. 2000;34(8): 1384-1389.

[42] Ready R, Navrud S. Benefit transfer – the quick, the dirty, and the ugly? Choices 2005;20(3):195-199.

[43] Boyd J. Economic Valuation, Ecosystem Services, and Conservation Strategy. 2011. Ecosystem Services Seminar Series. http://www.moore.org/materials/white-papers/Ecosystem-Services-Seminar-3-Valuation.pdf (accessed September 22, 2014).

Chapter 7

Modelling of Best Management Practices in Agricultural Areas

Youn Shik Park, Kyoung Jae Lim, Jae E. Yang and Ki-Sung Kim

Additional information is available at the end of the chapter

http://dx.doi.org/10.5772/59990

1. Introduction

Concern over water and pollutants influencing human health has been increasing in the last few decades. Nonpoint source (NPS) pollution, especially, led to water quality degradation in watersheds; therefore, water quality in streams or rivers has been made subject to government regulations (e.g., the Clean Water Act). Typically, watersheds are composed of various land uses; agricultural areas were a possible major source of phosphorus in a watershed [1,2]. Approximately 50% of NPS is from agricultural areas, since 52% of total nitrogen, 47% of total phosphorus, and 46% of sediment in U.S. streams are from agricultural areas [3]. A high concentration of total nitrogen came from agricultural areas in the watershed, and fertilization during cropping in an agricultural area led to a high concentration of high nitrate and orthophosphate [4,5]. Pollutants from three watersheds were compared [5]: agricultural watershed (95% agriculture and 5% urban), mixed watershed (43% agriculture and 57% urban), and urban watershed (1% agriculture and 99% urban) (Figure 1). Nitrate and soluble phosphorus concentrations in a stream were higher in the agricultural watershed, the other nutrients' concentrations (total suspended solids, turbidity, and pH) were higher in the urban watershed [5]. In addition, total phosphorus and ammonium concentrations did not display much difference by watershed types. They indicated that agricultural activity such as fertilization led to the higher nitrate and soluble phosphorus concentrations in the agricultural watershed; that the pollutant loads of nutrients and sediment were significantly variable by sites and land uses due to flow quantity; and that pollutant load quantification can be difficult since it varies by sites, land uses, season, and flow. Therefore, best management practices (BMPs) to reduce or manage pollutant loads in watershed have been studied in the last few decades [6–9]. In this chapter, recent research of BMPs in agricultural areas in various watersheds, with various optimization techniques and hydrologic models, were introduced to provide the rationale via which the researchers estimated the impact of BMPs in agricultural areas and to identify the processes by which the researchers optimized BMPs in watersheds.

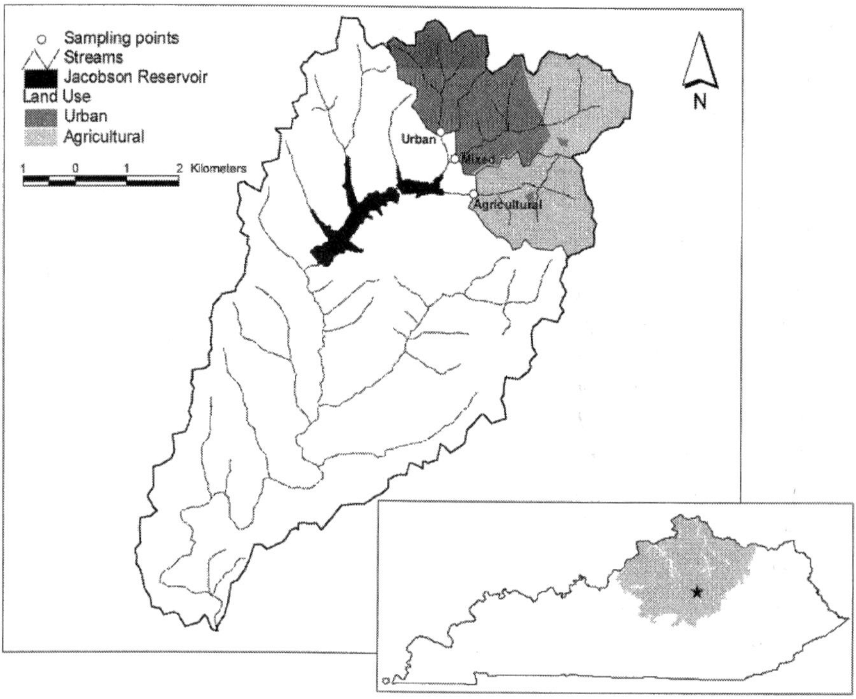

Figure 1. Agricultural, mixed, and urban watersheds in the State of Kentucky (adapted from [5])

2. Best management practices and pollutant control in watersheds

2.1. Best management practices in watersheds

BMPs were originated for soil erosion control [10] and have recently been implemented to control other NPS [11–14]. It can be readily found that BMPs were optimized using a hydrology model with either a straightforward or a sophisticated approach [15–18]. According to the study by Rao et al. [2], Variable Source Loading Function (VSLF) [19] was applied in a 164 ha agricultural area to estimate BMPs for dissolved phosphorus and total phosphorus. VSLF defines hydrologic response units (HRUs) by the soil wetness index and land uses; BMPs can be applied by pollutant reduction coefficients (or ratio). Three BMPs (crop rotation, strip cropping, and filter strip) for agricultural area were applied, and pollutant reduction coefficients were defined based on BMPs and soil wetness index. BMPs were implemented in the middle of a model simulation period, therefore the VSLF model was calibrated through the period prior to BMP implementation, and the pollutant reduction coefficients were calibrated through the period subsequent to BMP implementation. The study pointed out that BMPs led to decreases runoff losses.

A Soil and Water Assessment Tool (SWAT) [20] model was applied to a watershed of 50 km^2, and 55% of the watershed was an agricultural area [21]. The effect of four BMPs (crop rotation, reducing nutrient application, reduction of livestock density, and buffer strip) were estimated for annual total nitrogen, sediment, and total phosphorus. Annual total nitrogen reduction ranged from 9.9% to 46.7%, however, the BMPs were not effective at reducing sediment and total phosphorus due to the fact that annual sediment reduction by BMPs ranged from 0.82% to 11.9% and that annual phosphorus reduction by BMPs ranged from 1.1% to 13.6% (Table 1). Since both BMP effectiveness and BMP implementation costs are important, BMP implementation costs were analysed for the watershed based on establishment and maintenance costs. In the study, it was concluded that 1) crop rotation was not applicable since the implementation cost was extremely high, and 2) BMPs were not effective for annual sediment and total phosphorus reduction, while BMPs were effective for annual total nitrogen reduction.

BMPs	Total nitrogen	Sediment	Total phosphorus
Crop rotation	9.9%		4.6%
Reducing nutrient application	8.6%	0.8%	1.1%
Reduction of livestock density	15.6%	3.5%	3.9%
Filter strip	12.9%	4.9%	5.3%
Combined BMPs	46.7%	11.9%	13.6%

Table 1. Nutrient reductions by BMPs

There are several rationales to focus on regarding BMP implementations in agricultural areas. One is that the agricultural area is typically one of the major sources of NPS in a watershed. There are several research studies indicating that the pollutant loads from agricultural areas are more severe than those from urban areas in watersheds [5,22], since agricultural activities increase pollutant concentrations (or loads) during the cropping season [5,23], and the exposed soil surface has more soil erosion potential [24].

The other rationale to implement BMPs in an agricultural area is the cost. BMP implementation costs can be estimated by the Equations 1 [25] or 2 [26], for instance, which require an establishment cost (or capital cost). The costs for urban BMPs are typically more expensive than the costs for agricultural BMPs (Table 2). Reference [18] estimated the impact of BMPs for a watershed; the most cost-effective BMP was a filter strip for an agricultural area reducing 147.5 kg/year for $7,650, while the impact of a filter strip on urban areas was estimated to reduce 1 kg/year for $2,040. Therefore, the BMP implementation in agricultural area are typically more cost-effective than in urban.

$$A_{BMP} = \frac{Z \cdot (\frac{s}{100})}{1-(1+\frac{s}{100})^{-td}} \quad (1)$$

$$c_t = c_0 \cdot (1+s)^{td} + c_0 \cdot rm \cdot \left[\sum_{i=1}^{td} (1+s)^{(i-1)} \right] \quad (2)$$

Where A_{BMP} is the annual cost for a BMP, Z is the capital cost ($) of a BMP, *s* is the interest rate (%), *td* is the BMP design life, c_t is the BMP implementation cost ($/ha), c_0 is the establishment cost (%/ha), and *rm* is the ratio of annual maintenance cost to establishment cost (i.e., the percentage of establishment cost).

Land uses	BMP	C_0	*rm*	Reference
Cropland	Contour farming	15	1	[23]
	Filter strip	21	10	[24]
	Reduced tillage systems	7	1	[25]
Forest	Site preparation/hydro mulch/seed/fertilizer	3707	1	[26]
	Site preparation/straw/crimp/net	35,481	1	[27]
Feedlots	Filter strip	21	10	[24]
Urban	Alum treatment	1,112	0	[28]
	Grass swales	1,730	5	[29]
	Infiltration basin	7,413	3	[29]
	Infiltration trench	22,239	5	[29]
	Porous pavement	592,015	1	[30]
	Sand filter	25,946	12	[29]
	Vegetated filter strips	2,224	4	[29]
	Weekly street sweeping	14,947	7	[30]
	Wetland detention	6,178	2	[29]

Table 2. BMP costs for land uses [18]

2.2. Optimizations of best management practices

The previous sections introduced why BMP simulations were often applied in agricultural areas and how BMP implementation costs could be estimated, since BMP implementations at watershed are required to be cost-effective. In this section, several related research results are reviewed and presented to introduce how BMPs were optimized to be cost-effective BMP scenarios.

Gitau et al. [35] optimized three BMPs for an average annual loading reduction of dissolved phosphorus of 60%: which were contour strip cropping, having a nutrient management plan, and riparian forest buffers. To demonstrate BMP optimization, a farm of 300 ha was selected: which is located at the Town Brook watershed in New York State. Cropland and pasture areas are 44% and 19% in the study area, BMP implementation was considered for cropland and pasture areas. The BMP optimization process comprised four components. The first component was a SWAT model to simulate dissolved phosphorus and BMPs in the study area. BMPs were assigned for each HRU, since the model divides a watershed into sub-watersheds and HRUs defined by land use, hydrologic soil group, and slope. The second component was the BMP tool [36]. The BMP tool has a database of BMP effectiveness data for 32 BMPs, which were collected from published BMP monitoring studies. The BMP effectiveness database contains particulate phosphorus, dissolved phosphorus, total phosphorus, nitrogen, sediment, and runoff reduction by BMPs. The BMPs can be categorized into eight classes: animal waste systems, barnyard runoff management, conservation tillage, contour strip crop, crop rotation, vegetated filter strips, nutrient management plans, and riparian forest buffers. The third component was computing annual BMP implementation costs considering capital cost, interest rate, and BMP design life. Annual costs were $11/ha for contour strip cropping, $27/ha for nutrient management plan, and $1,942/ha for riparian forest buffers. Contour strip cropping was considered for cropland and pasture, a nutrient management plan for cropland and pasture, and riparian forest buffers for all agricultural areas bordering a stream. The last component was a genetic algorithm (GA). The study area was divided into 168 HRUs, and 149 HRUs which were croplands or pastures. The GA optimized the BMPs for the HRUs of cropland and pasture based on the annual implementation costs.

Two solution scenarios to implement BMPs were established (Figure 2). Both scenarios were able to reduce 60% of dissolved phosphorus at the cost of $1,430/year and $1,683/year, respectively. One of the scenarios required slightly lower costs, since the scenario applied BMPs to a smaller study area than the other scenario did. The authors mentioned that the BMP optimization technique they used is applicable for other studies to optimize BMPs on the level of average annual estimation. However, they also mentioned that their results were site-specific; therefore, the method cannot be used directly if land uses or site characteristics are different.

In the study of Maringanti et al. [37], two BMPs were optimized at Wildcat Creek Watershed located in Indiana, for atrazine reduction. Atrazine is one of the herbicides used in corn production and is used for broadleaf weed control. The watershed, the Wildcat Creek Watershed, is comprised of 74% of agricultural area, 21% of pasture, and 3% of urban. The watershed area is 1,956 km^2; corn is cultivated in 743 km^2 (38% of the entire watershed area), and soybeans are cultivated in 704 km^2 (36% of the entire watershed area). Atrazine led to water quality problems in Indiana, since the high atrazine level has degraded water quality in many watersheds. Therefore, the researchers studied BMP optimizations for atrazine reduction where it is used as a pesticide. The BMP optimization process of the study was also composed of four components. The first component was a SWAT model to estimate pesticide concentration, and BMPs that were buffer strips and tillage practices. The watershed was divided into 109 sub-basins, and 403 HRUs were identified in the watershed. The second component was BMP effectiveness (Table 3).

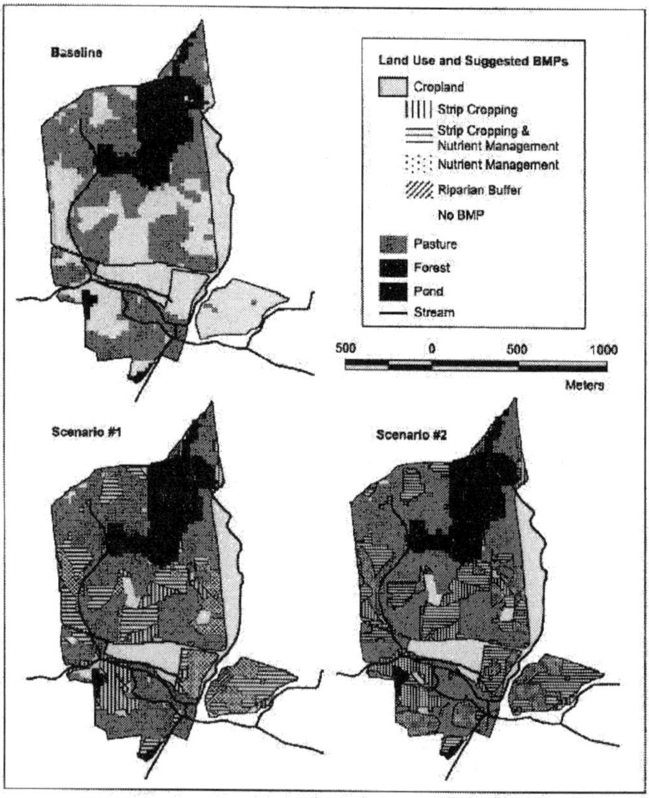

Figure 2. Optimized BMPs for the Cannonsville Reservoir watershed in New York State [35]

BMPs		Pesticide reduction	Net cost increase
Buffer width	Tillage practice	%	$/ha
0 m	Conventional	-	-
0 m	No-till	7.1	3
20 m	Conventional	41.9	245
20 m	No-till	41.4	242
27 m	Conventional	45.7	327
27 m	No-till	44.4	324
30 m	Conventional	46.9	409
30 m	No-till	45.4	406

Table 3. Pesticide reduction effectiveness and net costs

Regarding BMP simulations and optimizations, the researchers made five assumptions. The first assumption was that the BMP effectiveness in terms of HRU levels in the SWAT model is identical to (or only marginally different from) the BMP effectiveness in watershed level. The assumption was derived from the SWAT model characteristic that the model computes hydrology at an HRU level and that the model is watershed scale model. The second assumption was that the effectiveness of BMPs does not vary temporally or seasonally. For instance, the pollutant reduction by buffer strips might vary based on the growth (or height) of vegetation in buffer strips; however, it was limited to considering the variance of BMPs. The third assumption was that only atrazine pesticide was considered, since the SWAT model does not allow consideration of multiple pesticides, whereas multiple pesticides may in fact be applied in the fields. The fourth assumption was that stream routing processes can be disregarded. The fifth assumption was that BMPs are watershed-specific.

The third component was implementation costs using Equation 2. The interest rate was 6.5%, maintenance rates were 1% for buffer strips, and the design life for buffer strips was 10 years. The fourth component was the multi-objective GA. The 403 HRUs were the variables to be determined, satisfying the two objective functions of minimizing pollutant loading and net cost increases.

Two solution scenarios to implement BMPs were found. Figure 3 (c) displays the base scenario that does not contain any BMPs. The scenario shown in Figure 3 (a) required a net cost increase of $97/ha, and the scenario shown as Figure 3 (b) required a net cost increase of $35/ha in the study area. The study indicated that the BMP optimization by a multi-objective GA performed well; however, the approach using a multi-objective GA may rule out some solutions, an inherent drawback of the multi-objective GA.

Similar to the case study discussed above, Maringanti et al. [38] optimized BMPs in the Wildcat Creek Watershed, Indiana State. An identical model to simulate pollutants and an identical optimization technique to optimize BMP implementation plans were used, which were SWAT and a multi-objective GA. However, BMP implementation plans were to reduce nitrogen, phosphorus, sediment, and pesticide. Eight BMP combinations were prepared for 2 BMPs in the previous study. [37]; however, 160 BMP combinations were prepared for a HRU with multiple BMPs [38]. The 160 BMP combinations were developed from the conditions of five filter strips, two contour farming, four residue managements, two parallel terraces, and two tillage types (Table 4).

Distinct features were found in the study. The first feature was that contour farming without a filter strip had a negative impact by increasing the total nitrogen and total phosphorus. The second feature was that a filter strip with a parallel terrace was very effective for pollutant reduction. Residue management did not provide pollutant reduction, and contour farming was effective only for pesticide reduction. No-till was effective for pesticide reduction, but did not reduce total phosphorus.

The researchers investigated four pollutants (nitrogen, phosphorus, sediment, and pesticide) and cost; therefore, they were able to summarize their results in a spider plot spatially representing different costs for BMPs in the watershed (Figure 4 and 5). The numbers in Figure

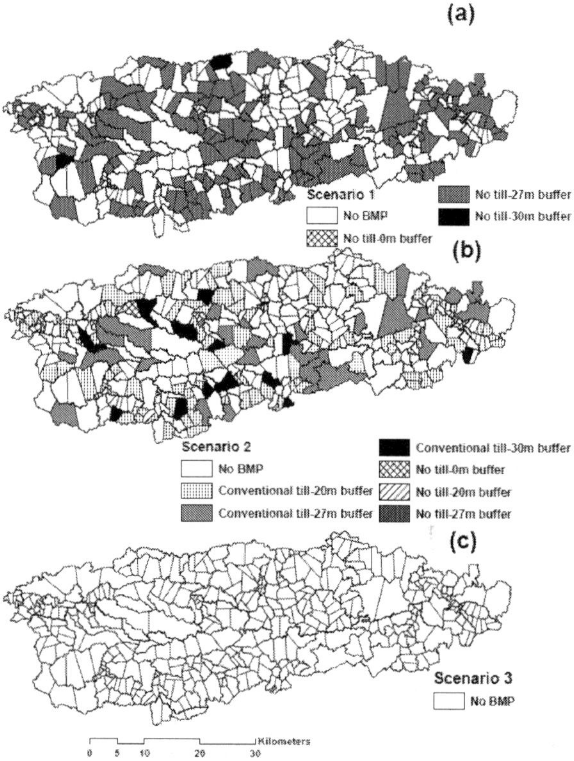

Figure 3. Optimized BMPs for the Wildcat Creek Watershed in Indiana State [37]

5 are the number of BMP combinations: for instance, BMP combination number 1 was comprised of 'residue management of 1,000 kg/ha' and 'conservation till', number 33 was comprised of a 'filter strip of 10 m'.

BMP	BMP Type	Cost	Unit
Filter strips	0, 5, 10, 20, 30	12.2	$/ha/m
Contour farming	Not present or present	16.8	$/ha
Residue management	1000, 3000, 5000, 7000 kg/ha	0.0	$/ha
Parallel terrace	Not present or present	74.9	$/ha
Tillage	Conservational, No-till	53.1	$/ha

Table 4. BMPs and BMP implementation costs [38]

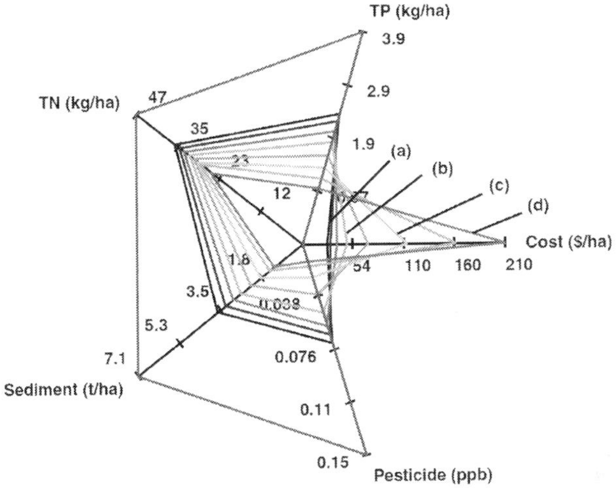

Figure 4. Spider plot representing the different nonpoint source pollutant loads after the final generation by multi-objective GA [38]

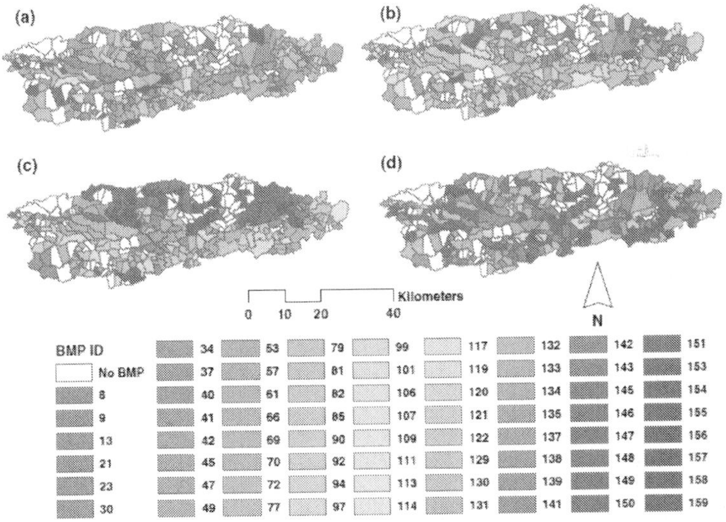

Figure 5. Locations and types of optimized BMPs in the Wildcat Creek Watershed [38]

Veith et al. [1] optimized BMP locations in a watershed of 1,014 ha located in the State of Virginia. An approach using the universal soil loss equation (USLE) with a sediment delivery ratio was used to estimate sediment loads, and GA was used to optimize BMP locations in the study area. BMPs were, of course, optimized by pollution reduction and BMP implementation costs. However, distinctive feature of the study was that the variance of crop production by BMP implementations was considered by two additional criteria. One was that the preference of feed production and nutrient management requirement was applied to the farms; the other was that it avoided applying BMPs to a few farms.

In the study performed by Srivastava et al. [39], two BMP optimization processes were compared which used design storm and continuous climate data with a model to estimate pollutant loads. An Annualized Agricultural Nonpoint Source pollution model (AnnAGNPS) [40] was used to estimate sediment, sediment nitrogen, dissolved nitrogen, sediment organic carbon, and sediment phosphorus in a watershed of 725 ha located in Northumberland County, Pennsylvania. The study area was comprised 47% cropland. Fifteen current crop rotations were considered, two design storms (69.85 mm as 2 year return period storm event, and 88.90 mm as 5 year return period storm event), and climate data for five years were used for BMP optimizations, since the hypothesis of the study was that a BMP scheme with an optimization process using accumulated pollutant loads from continuous simulation would be more applicable than the process using pollutant loads from several critical (or extreme) storms. Based on the results supporting the hypotheses, the authors suggested that long-term pollutant loads from continuous simulations need to be considered in a BMP optimization analyses.

A simple technique to optimize BMPs in a watershed was used by Park et al. [18]. They optimized BMPs for a watershed of 129.1 km^2, based on annual BMP implementation costs computed by Equation 1. Sophisticated techniques (e.g., a genetic algorithm) are often used in BMP optimization processes. The researchers, however, performed the optimization process by a straightforward approach using BMP implementation costs for unit mass reduction (or cost per 1 kg reduction of pollutant). The study had two applications with potential area for BMPs. One was that it is possible to apply a filter strip of up to 100% of the agricultural area (79 km^2), the other was that it is possible to apply a filter strip on up to 10 km^2 and reduce tillage systems in up to 10 km^2 of agricultural area. In the first application, pollutant reduction met the requirement for application of 17 km^2 filter strip at an estimated annual cost of $12,870. In the alternate application, the estimated annual cost was $17,400, which resulted from $7,650 for 10 km^2 of a filter strip in the agricultural area, $7,710 for 10 km^2 of reduced tillage system in the agricultural area, and $2,040 for 4 km^2 of vegetative filter strip in the urban area. The applications were to demonstrate the fact that BMP scenarios and implementation costs can vary by watershed conditions.

In this section, several recent research studies optimizing BMPs were introduced. Optimization techniques are complex but are used widely to solve problems; GAs, for example, have been applied in BMP optimizations with various hydrologic models. On the other hand, BMP optimizations have been performed by adopting a straightforward approach based on BMP implementation costs for unit mass reduction. Moreover, various hydrologic models were used to predict pollutant loadings and the impact of BMPs on watersheds. Complexity in optimization techniques or hydrology models was not crucial for BMP optimization processes. However, it was found that BMP optimization processes typically required four components regardless of which hydrology models and optimization techniques were selected. In other words, the processes were composed of 'selecting available BMPs for the watershed', 'gathering and computing annual BMP implementation costs', 'identifying optimization technique', and 'selecting a model to estimate pollutant loads and the impact of BMPs'.

3. Summary and discussion

NPS pollution has caused water quality degradation in streams and rivers, therefore, various research projects were concerned to perform NPS reduction. Research indicates that agricultural areas were typically major source of NPS in watersheds; therefore, there is a need to perform BMP implementations. Models (or computer software) and sophisticated techniques were often used to suggest optimized BMPs for watersheds. It can be stated that the BMP optimization processes are typically composed of four components: selecting available BMPs for the watershed (or site), gathering and computing annual BMP implementation costs, identifying optimization technique, and selecting a model to estimate pollutant loads and the impact of BMPs. The first two components are site-specific, since some BMPs have a limited application in certain watersheds, and BMP costs vary by location. As the researchers mentioned, BMPs optimized in other research studies cannot be selected and implemented identically without consideration of regional characteristics. Pollutant behaviours can differ by the locations of the source area in a watershed; thus, the locations of BMPs would be one of the important factors to consider. To summarize, BMP optimization processes should answer the following questions: 'what BMPs need to be selected?', 'what size of BMP needs to be applied?', 'where do BMPs need to be placed?', and 'how much does it cost to implement BMPs?'. The process will be very complex and will require a lot of effort; an optimization technique would therefore be required to examine varying BMP impacts, and this is why optimization techniques are often employed. Although optimization techniques provide convenience in BMP evaluations, recognizing limitations in hydrology models and optimization techniques is still required.

Acknowledgements

This research was supported by the Geo-Advanced Innovative Action (GAIA) Project (No. RE201402074, Surface Soil Resources Inventory & Integration: SSORII Research Group) in Republic of Korea.

Author details

Youn Shik Park[1], Kyoung Jae Lim[2*], Jae E. Yang[3] and Ki-Sung Kim[2]

*Address all correspondence to: kjlim@kangwon.ac.kr

1 Dept. of Rural Construction Engineering, Kongju National University, Chungcheongnam-do, South-Korea

2 Dept. of Regional Infrastructure Engineering, Kangwon National University, Kangwon-do, South-Korea

3 Dept. of Biological Environment, Kangwon National University, Kangwon-do, South-Korea

References

[1] Veith TL, Wolfe ML, Heatwole D. Optimization procedure for cost effective BMP placement at a watershed scale. Journal of the American Water Resources Association 2003; 39(6) 1331–1343.

[2] Rao NS, Easton ZM, Schneiderman EM, Zion MS, Lee DR, Steenhuis TS. Modeling watershed-scale effectiveness of agricultural best management practices to reduce phosphorus loading. Journal of Environmental Management 2009; 90 1385–1395.

[3] Allan JD. Stream Ecology, Structure and function of running waters: Springer Science and Business Media; 1995.

[4] McMahon G, Harned DA. Effect of environmental setting on sediment, nitrogen, and phosphorus concentrations in Albemarle-Pamlico Drainage Basin, North Carolina and Virginia, USA. Environmental Management 1998; 22 887–903.

[5] Coulter CB, Kolka RK, Thompson JA. Water quality in agricultural, urban, and mixed land use watersheds. Journal of the American Water Resources Association 2004; 1593–1601.

[6] Lee JW, Eom JS, Kim B, Jang WS, Ryu J, Kang H, Kim KS, Lim KJ. Water quality prediction at Mandae watershed using SWAT and water quality improvement with vegetative filter strip. Journal of the Korean Society of Agricultural Engineers 2011; 53(1) 1–9.

[7] Ryu J, Kang H, Kim NW, Jang WS, Lee KW, Moon JP, Lee KS, Lim KJ. Analysis of total nitrogen reduction efficiency with established riparian buffer system using SWAT-REMM model in Bonggok watershed. Journal of Korean Society of Water Quality 2010; 26(6) 910–918.

[8] Lee JM, Ryu J, Kang H, Kang H, Kum D, Jang CH, Choi JD, Lim KJ. Evaluation of SWAT flow and sediment and effects of soil erosion best management practices. Journal of the Korean Society of Agricultural Engineers 2012; 53(1) 99–108.

[9] Kum D, Jang CH, Shin MH, Choi J, Kim B, Jeong GC, Won CH, Lim KJ. Determination of model parameters of surface cover materials in evaluation of sediment reduction and its effects at watershed scale using SWAT. Journal of Korean Society of Water Quality 2012; 28(6) 923–932.

[10] Walter MF, Steenhuis TS, Haith DA. Nonpoint source pollution control by soil and water conservation practices. Transaction of ASAE 1979; 22 834–840.

[11] Brannan KM, Mostaghimi S, McClellan PW, Inamdar S. Animal waste BMP impacts on sediment and nutrient losses in runoff from the Owl Run watershed. Transactions of the ASAE 2000; 43 (5) 1155–1166.

[12] Lee K, Isenhart TM, Schultz RC, Mickelson SK. Multispecies riparian buffers trap sediment and nutrients during rainfall simulations. Journal of Environmental Quality 2000; 29(4) 1200–1205.

[13] Kim YJ, Geohring LD, Jeon JH, Collick AS, Giri SK, Steenhuis TS. Evaluation of the effectiveness of vegetative filter strips for phosphorus removal with use of a tracer. Journal of Soil Water Conservation 2006; 61 (5) 293–303.

[14] Bishop PL, Hively WD, Stedinger JR, Rafferty MR, Lojpersberger JL, Bloomfield JA. Multivariate analysis of paired watershed data to evaluate agricultural best management practice effects on stream water phosphorus. Journal of Environmental Quality 2005: 34 (3) 1087–1101.

[15] Srivastava P, Hamlett JM, Rovillard PD. Watershed optimization of agricultural best management practices: continuous simulation versus design storms. Journal of the American Water Resources Association 2003; 39(5) 1043–1054.

[16] Maringanti C, Chaubey I, Popp J. Development of a multiobjective optimization tool for the selection and placement of best management practices for nonpoint source pollution control. Water Resources Research 2009; 45 doi:10.1029/2008WR007094.

[17] Lee JG, Selvakumar A, Alvi K, Riverson J, Zhen JX, Shoemaker L, Lai F. A watershed-scale design optimization model for stormwater best management practices. Environmental Modeling and Software 2012; 37 6–18.

[18] Park YS, Engel BA, Harbor J. A web-based model to estimate the impact of best management practices. Water 2014; 6 455–471.

[19] Schneiderman EM, Steenhuis TS, Thongs DJ, Easton ZM, Zion MS, Mendoza GF, Walter MT, Neal AC. Incorporating variable source area hydrology into curve number based watershed loading functions. Hydrological Process 2007; 21(25) 3420–3430.

[20] Arnold JG, Srinivasan R, Muttiah RS, Williams JR. Large area hydrologic modeling and assessment part I: model development. Journal of the American Water Resources Association 1998; 34 73–89.

[21] Lam QD, Schmalz B, Fohrer N. The impact of agricultural best management practices on water quality in a North German lowland catchment. Environmental Monitoring and Assessment 2011; 183 351–379.

[22] Tetra Tech, Inc. User's Guide Spreadsheet tool for the estimation of pollutant load (STEPL) version 4.1.: Tetra Tech, Inc.; 2011.

[23] González-Estrada E, Rodriguez LC, Walen VK, Naab JB, Koo J, Jones JW, Herrero M, Thornton PK. Carbon sequestration and farm income in West Africa: Identifying best management practices for smallholder agricultural systems in northern Ghana. Ecological Economics 2008; 67 492–502.

[24] El-Hassanin AS, Labib TM, Gaber EI. Effect of vegetation cover and land slope on runoff and soil losses from the watersheds of Burundi. Agriculture, Ecosystems and Environment 1993; 43: 301–308.

[25] Degarmo EP, Sullivan WG, Bontadelli JA, Wicks EM. Engineering Economy: Prentice Hall International Paperback Editions; 1997.

[26] Arabi M, Govindaraju RS, Hantush MM. Cost-effective allocation of watershed management practices using a genetic algorithm. Water Resources Research 2006; 42 W10429 DOI: 10.1029/2006WR004931.

[27] Pertsova CC. Ecological Economics Research Trends: Nova Publishers; 2007

[28] Buckner ER. An evaluation of the use of vegetative filter strips on agricultural lands in the Upper Wabash River Basin. PhD thesis. Purdue University; 2001.

[29] Kieser and Associates. Modeling of agricultural BMP scenarios in the Paw River Watershed using the Soil and Water Assessment Tool (SWAT): Kieser and Associates; 2007.

[30] United States Environmental Protection Agency. National management measures to control nonpoint source pollution from forestry: United States Environmental Protection Agency; 2005.

[31] Great Lakes Environmental Center (GLEC). National level assessment of water quality impairments related to forest roads and their prevention by best management practices: Great Lakes Environmental Center; 2008.

[32] Wisconsin Department of Natural Resources. Alum treatments to control phosphorus in lakes: Wisconsin Department of Natural Resources, Madison, Wisconsin; 2003.

[33] United States Environmental Protection Agency. Preliminary data summary of urban storm water best management practices: United States Environmental Protection Agency; 1999.

[34] King D, Hagan P. Costs of stormwater management practices in Maryland Counties: University of Maryland Center for Environmental Science; 2011.

[35] Gitau MW, Veith TL, Gburek WJ. Farm-level optimization of BMP placement for cost-effective pollution reduction. Transactions of the ASAE 2004; 47(6) 1923–1931.

[36] Gitau MW. A quantitative assessment of BMP effectiveness for phosphorus pollution control: The Town Brook Watershed. PhD thesis. University Park; 2003.

[37] Maringanti C, Chaubey I, Arabi M, Engel B. A multi-objective optimization tool for the selection and placement of BMPs for pesticide control. Hydrology and Earth System Sciences Discussions 2008; 5 1821–1862.

[38] Maringanti C, Chaubey I, Arabi M, Engel B. Application of a multi-objective optimization method to provide least cost alternatives for NPS pollution control. Environmental Management 2011; 48 448–461.

[39] Srivastava P, Hamlett JM, Robillard PD. Watershed optimization of agricultural best management practices: continuous simulation versus design storms. Journal of the American Water Resources Association 2003; 1043–1054.

[40] National Sedimentation Laboratory. AnnAGNPS User Documentation: U.S. Department of Agriculture; 1998.

Chapter 8

Effluent Cleaning, Greener Catalysts and Bioecomaterials from Agricultural Wastes

A.M. Martínez Serrano, M. Ramos, M. Yates, M.A. Martin-Luengo, F. Plou, J.L. Lacomba, G. Reilly, C. Vervaet, P. Muñoz, G. Garcia, J.L. Tarterra, B. Fite, A. Urtzainki, M.C. Vidal, E. Sáez Rojo, L. Vega Argomaniz, A. Civantos and V. Zurdo

Additional information is available at the end of the chapter

http://dx.doi.org/10.5772/60018

1. Introduction

The transformation of subproducts and residues is of utmost importance, especially when it closes an industrial cycle, where solutions for environmental problems such as contamination leads towards a sustainable development, converting residues into value added products. Materials prepared from agricultural residues may be considered as Renewable Raw Materials [1,2]. Using these materials as a source for useful materials avoids the expense of employing other materials that are often non-renewable. This philosophy is denoted as "cradle to grave", since the residues of a company are used by the same or others [3].

Agriculture is one of the pillars of society, especially given the increasing world population. Countries which have large agricultural resources, such as Spain, also produce vast amounts of residues and wastes, that can be an environmental hazard, difficult to store, easily generating leakages and greenhouse gases such as ammonia, methane and carbon, nitrogen and sulphur oxides when burnt, therefore negatively impacting on the environment. However, these agriwastes can also be considered as a source of low cost renewable raw materials (RRM) that with the proper treatment can have applications in a wide range of processes, *i.e.* energy production, materials (fertilizers, animal feed, biodegradable plastics, resins, textiles, fibres, paper, *etc.*) and chemicals (platform molecules, solvents, additives), inside the Biorefinery concept, with the added bonus of not competing with food resources [4].

The work presented here uses agricultural wastes to prepare materials for effluent cleaning with rice and beer production wastes, etherification of glycerol from biodiesel production, enzyme immobilisation and use in biodiesel preparation and Bioecomaterials for tissue engineering and controlled desorption of bioactive substances.

2. Effluent cleaning with materials derived from rice and beer production wastes

Industrial effluents often have high flows with variable concentrations of toxic substances, where adsorption is an ideal process for their decontamination due to its good economic turnover and the possibility of recovering the contaminants by desorption. The process of adsorption is based on the diffusion of sorbates to the adsorbent surface, followed by the inclusion inside the pore structure where they are stored. Here the textural properties of the adsorbents (surface area and porosity) are of the utmost importance [5].

Distilleries Muñoz Galvez (DMG) is a leading Spanish Company that manufactures and exports Essential Oils and Aromatic Raw Materials (Fine Chemicals) as well as Fragrances and Flavours worldwide [6]. The challenge faced by the research group (CDTI project) [7] was to clean DMG wastewaters that contained terpenes as the main residue, as expected given their origin, thus reducing water consumption, economic expenditure and possible environmental hazards. Aiming to decrease the economic costs, and increase the project's environmental approach, agriresidues from beer and rice production were used to prepare materials capable of wastewater decontamination that were compared to commercial adsorbents. Previous knowledge of the researchers was applied and optimised to the wastewater composition and treatments [8,9]. The results indicated better adsorption and therefore cleaning capacities in the residue derived materials than in the commercial ones. The chemical oxygen demands after wastewater treatment were low enough for the treated water to be discharged in accordance with the legal requirements.

The agriresidues chosen for this study were beer bagasse and rice husk. Beer bagasse, from Mahou San Miguel (Spain) [10] (designated as BBM), is a residue from beer production that has *ca.* 75-80% water. Previous work by the authors has demonstrated that materials prepared from beer bagasse are basic due to their high P, Si, Ca and Mg content. Furthermore, due to their origin these materials are competitive in price; in fact BBM is at present used as a fertiliser [11].

Rice husk (RH) from DACSA (Spain) [12] is an agriresidue from rice production, of difficult storage and transport due to its high volume to weight ratio. On calcining this residue, a material with more than 97% silica is produced, that has also low amounts of calcium and potassium [13-15].

The agriresidues utilized as raw materials in this study were used as received, in the case of rice husk, or after a drying step at 150°C in the case of the BBM, to inhibit further fermentation due to its high water content. The RH and the dried BBM were analysed by TG-DTA in air to determine their thermal stabilities and design the material preparation. Details of experimental set up can be found in [14]. Figure 1 shows these data.

http://dx.doi.org/10.5772/60018

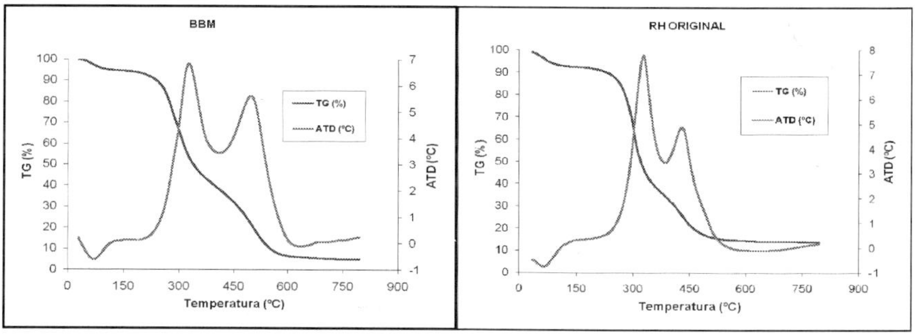

Figure 1. TG-DTA analyses of RH and BBM agriresidues in air.

From these results the agriresidues loose up to 90 (BBM) and 87 (RH) percent weight, when calcined up to 800°C, where the weight losses were an endotherm due to water loss up to 200 °C and two exotherms at 340 °C and 520 °C (BBM) or 340 °C and 470 °C (RH) due to decomposition/transformation of the organic matter to volatile organic substances and carbonization [14].

After studying the reproducibility of the materials the temperature chosen to prepare the adsorbents through thermal treatment of the agriresidues was 350 °C for 2, 4 or 6 h. These materials were designated as BBM2, BBM4, BBM6, RH2, RH4 and RH6. The composition analyses of these agriresidue derived materials are included in Table 1. From these results it was clear that under these conditions the materials derived from BBM were mainly carbonaceous whilst those derived from RH have siliceous structure and the content of the elements that form volatile species (CO_2, H_2O, NO_X, oxychlorides, *etc.*) decreased on increasing the time of thermal treatment, while the percentage of the other elements increased.

	BBM2	BBM4	BBM6	RH2	RH4	RH6
C	43.1	43.2	40.1	7.0	1.8	1.0
H	1.6	1.6	1.6	0.7	0.4	0.3
N	11.3	10.2	2.5	0.5	0.2	0.1
Al	0.3	0.2	0.0	0.0	0.0	0.0
Na	0.1	0.1	0.1	0.1	0.1	0.2
Mg	1.2	1.6	1.7	0.4	0.4	0.4
Si	2.1	5.6	7.9	42.1	52.0	55.6
P	1.5	3.5	5.6	0.4	0.5	0.5
S	0.2	0.2	0.2	0.2	0.2	0.2
Cl	0.0	0.0	0.1	0.0	0.0	0.1
K	0.2	0.4	0.6	2.3	3.4	3.6
Ca	0.9	1.6	2.5	0.9	1.1	1.1
Fe	0.1	0.1	0.2	0.1	0.1	0.2
Zn	0.0	0.0	0.1	0.0	0.0	0.1

Table 1. Percent composition of agriresidue derived materials (Traces: Ti, Cr, Ni, Ga, Br, Rb, Mn, Sr, Cu).

Textural analyses of the agriresidue derived materials were carried out by mercury intrusion porosimetry (MIP) and the data obtained are included in Figure 2 and Table 2. Description by the authors of the experimental analyses for MIP can be found in reference 16.

In this technique, the pores below 300 nm correspond to those inside the particles, while those at higher values are due to interparticle voids. As can be seen in all the materials prepared, the curves coincide for values below 1000 nm. As expected, in general, there was an increase in the total pore volume on increasing the calcination time, due to decomposition of volatile compounds, producing extra porosity, in agreement with TG-DTA analyses.

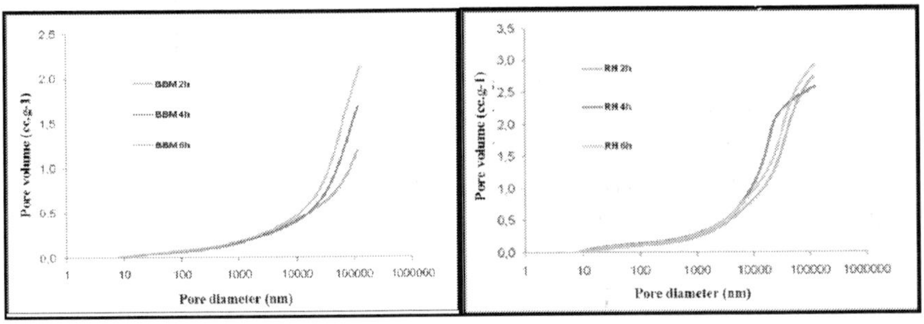

Figure 2. MIP textural data of agriresidue derived materials.

	S_{Hg} (m²/g)	Vp (cm³/g)	Porosity (%)	dp (nm)
BBM2	16	0.18	66.7	25.4
BBM4	18	0.18	69.2	44.7
BBM6	20	0.19	75.9	67.1
RH2	27	0.24	81.0	36.1
RH4	33	0.25	82.3	48.3
RH6	38	0.26	84.7	67.9
Fluesorb B	52	0.22	44.6	10.2

Table 2. Textural characteristics of agriresidue derived materials and an Activated Carbon by MIP. *(S_{Hg}= Specific surace area, Vp = Pore Volume, dp = medium pore size.*

From this data it can be observed that the porosity, surface area and medium pore size increase with time of calcination and the surface areas were lower but the pore sizes higher for the residue derived materials compared to the commercial activated carbon.

The analysis of DMG wastewater was carried out by GCMS (Figure 3), the experimental details for this technique can be found in reference 17. From these analyses the presence of *ca.* 16 different substances was found (Figure 3) and for the quantification of the cleaning studies, the main substances (9.5 min (pinene), 13.6 min (limonene), 18.1 min (hexadecyl acetate), 19.5min ($C_{19}H_{40}$), 22 min ($C_{20}H_{42}$), 26.2 min ($C_{21}H_{44}$) and 27.5 min ($C_{22}H_{46}$) were chosen.

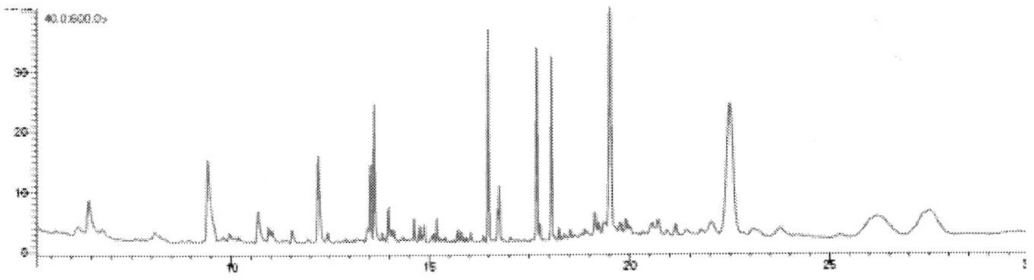

Figure 3. GCMS analysis of DMG wastewater.

The DMG wastewater treatments were carried out by using 0.5 g of carbon (commercial or Ecomaterial), and 30 mL of the DMG wastewater (to ensure reproducibility of the measurements), these slurries were magnetically stirred and 2 mL aliquots of the original and treated wastewaters were extracted at increasing times with the same volume of di-isopropylether, dried over sodium sulphate and analysed by GCMS (Figures 4 and 5).

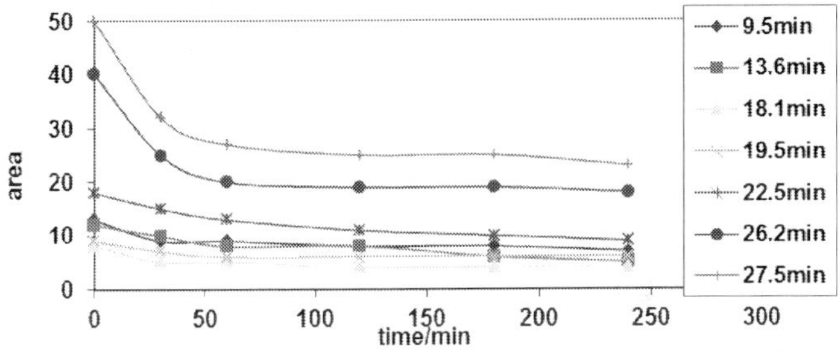

Figure 4. DMG Wastewater cleaning on BBM2.

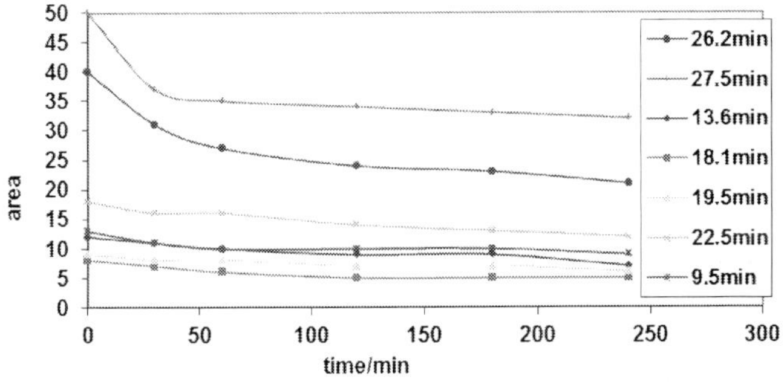

Figure 5. DMG Wastewater cleaning on Fluesorb B.

These results indicated that the carbon prepared from beer residues had better adsorption capacities for cleaning the DMG wastewater than the commercial one, especially with regards to the high molecular weight substances. This can be related to the different textures, since the commercial carbon has smaller pores that can not easily accommodate the high molecular weight substances. The chemical oxygen demand (COD) of the original and treated wastewaters was studied according to the Spanish UNE 77004, equivalent to ISO 6060:1989:

$$\mathrm{DQO}: \frac{800 \times C \times (V_1 - V_2)}{V_0}$$

where C: concentration of Fe(II) sulphate and ammoniumin mol/L, V_0: volume in mL before dilution, V_1: volume in mL of Fe(II) sulphate and ammonium solution for blank analysis, V_2: is the volume in mL of Fe(II) sulphate and ammonium solution for assay, 8000 is molar mass in mg/L of ½ O_2. The COD results are in mg O_2/L. The value of the method has been checked with a 0.425 g of potassium hydrogenphtalate ($KC_8H_5O_4$), dried at 105°C, diluted in 1000 mL distilled water with a COD standard value of 500 mg O_2/L (+/- 20). Variabilities in COD analyses were less than 2 % [18]. The results obtained for the COD reduction of the wastewater, with the different materials are included in Table 3 (Percent reduction of COD after room temperature wastewater treatment with adsorbents until constant COD (usually *ca.* 60 minutes).

	COD **% Reduction**
Comercial (Fluesorb B)	**4**
BBM2	5
BBM4	10
BBM6	18
RH2	17
RH4	67
RH6	73

Table 3. Percent COD reduction of DMG wastewater with adsorbents.

The high effectiveness of the residue derived materials compared with the commercial carbon should be noted. The material with greater cleaning ability was RH6 (73% reduction), which allowed a water with COD of 960 mg O_2 /L. Comparing the textural data with COD determinations, it can be said that there was a direct correlation between the pore diameter of the solids and their COD reduction capacity. Thus, wastewater treatment with residue derived materials has been shown to be an economical and environmentally sound process that should be further developed [19].

3. Etherification of glycerol from biodiesel production with company's own residues

The need for renewable energies in general and biodiesel in particular, indicates that optimising the production process is of vital importance. Biodiesel production generates *ca.* 10% of glycerol as a subproduct which has led to a fall in the glycerol prices, making the search for other industrial applications a neccesity. Amongst all possible processes to increase the value of glycerol, etherification is one of the most promising, since glycerol ethers can be used as such or with slight modification as fuel additives [20]. Other important uses are found in cosmetics, food additives, monomers for polimerisation processes *etc.* [21].

This research was undertaken with the aim to use Acesur-Tarancon's own subproducts to prepare catalysts to transform the company's glycerol (from their biodiesel plant) to diglicerol ethers. The catalysts prepared in this way are in fact Ecomaterials and their origin makes them competitive with commercial ones. Production of ethers with more than three glycerol molecules competes with diethers and thus control of the selectitivy is important, especially keeping in mind that the diglycerol ethers (Figure 6) are produced at short reaction times, which gives the process an added value [20, 22].

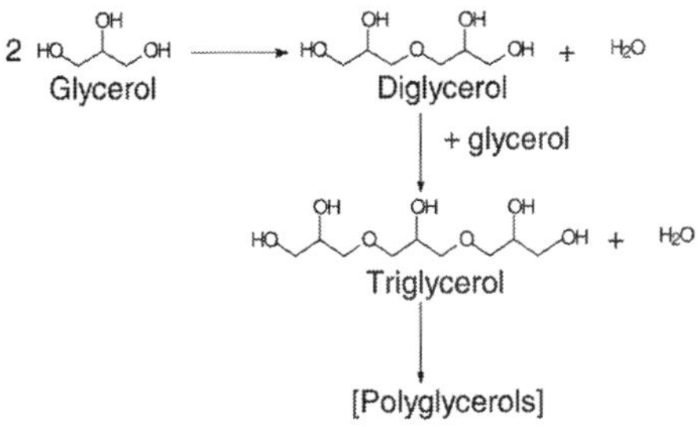

Figure 6. Glycerol pathway to di- and triglycerols in reference 23.

Etherification of glycerol with acid catalysts was found to be difficult to control. However, the catalytic transformation of glycerol into ethers, carried out with basic catalysts allows more controllable results. Furthermore, the use of heterogeneous catalysts *i.e.* alkaline and alkaline earth oxides, present in the residue derived materials form Acesur, compared to homogeneous bases is gaining interest since they are easily separated from the reactants and products for reuse with the corresponding economic benefits [24]. A bibliographic search showed Barrault 's work describing caesium oxide catalysts that achieved medium conversions with selectivities to di- and triglycerols, depending mainly on the reaction time [25]. Also Ruppert describes a reaction carried out on alkaline-earth oxides at 220 °C for 20 h giving rise to higher glycerol conversions on the more basic catalysts: 5 % (MgO), 58 % (CaO), 80 % (SrO) and 80 % (BaO) [26].

The oligoglycerols are gaining more and more interest as products used in cosmetics, food-additives or lubricants [27]. Short overviews about the synthesis of glycerol oligomers from di- to pentaglycerol have been published by Rollin *et al* [28]. Generally oligoglycerols are produced using basic homogeneous catalysis, but lately increased attention has been paid towards the heterogeneously catalysed processes. Despite a lower activity heterogeneous catalysts reveal many advantages: firstly, the separation of the catalyst and secondly, by carrying out the reaction in the absence of solvent, in this work only filtering the catalyst was needed, with evident economic and environmental advantages.

The conditions used for glycerol etherification were chosen with basic catalysis, since the sunflower oil production agriresidue derived materials (RP45), given their composition of alkaline (26 % potassium) and alkaline-earth cations (5 % magnesium and 7 % calcium) are basic in nature. TGMS of adsorbed acetic acid indicated that RP45 contained basic centres of low (100-200 °C), medium (200-500 °C) and high basicity (550-650 °C) (Figure 7) (see procedure for basicity measurement in reference 14) and can catalyse Knoevenagel condensation reactions [29].

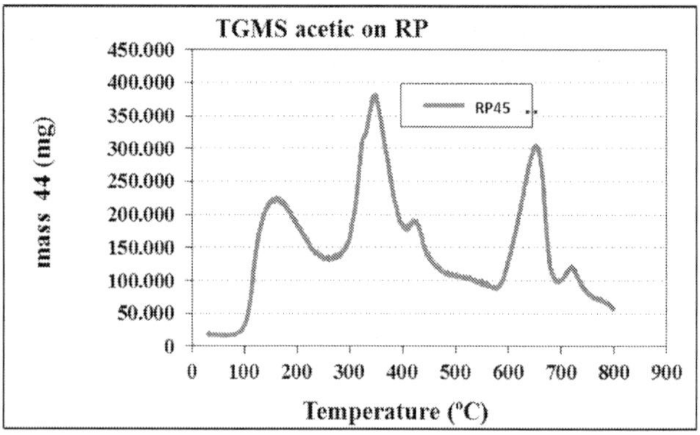

Figure 7. Analysis of basicity of RP45 by TG-MS of acetic acid decomposition.

The reactions were carried out in abscense of solvent, under inert atmosphere to limit over-oxidation, controlling temperature and time of reaction to optimise the economics [30, 31]. The work described here was carried out to produce diglycerol ethers in the absence of solvent, under inert atmosphere to avoid overoxidation, using catalysts derived from sunflower oil production residues, which have medium and high basic strengths. The analysis of reaction products was carried out by GC-MS (conditions as in reference 14).

Optimisation of the reaction conditions (Figure 8) showed that with 240°C, under inert atmosphere (nitrogen flow) and a ratio catalyst/glycerol = 1/50, after 4 h the conversions of glycerol and selectivities to diglycerols were optimum. At lower temperatures the conversions were low and with higher temperatures the selectivity to diglycerols decreased due mainly to uncontrolled formation of polyglycerols and oxidised compounds (mainly glycolic and glyceric acids).

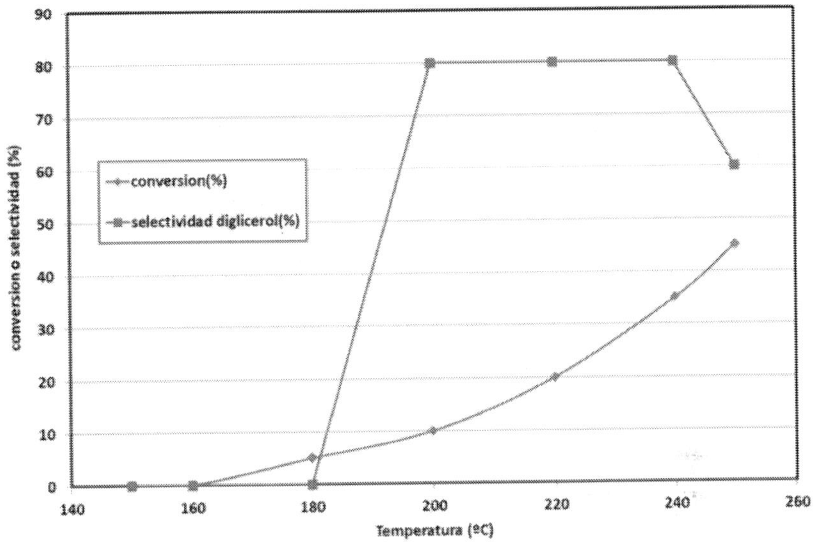

Figure 8. Conversions and selectivities of glycerol to diglicerol at different reaction temperatures.

Main compounds found in the reaction carried out in this work are summarised in Figure 9.

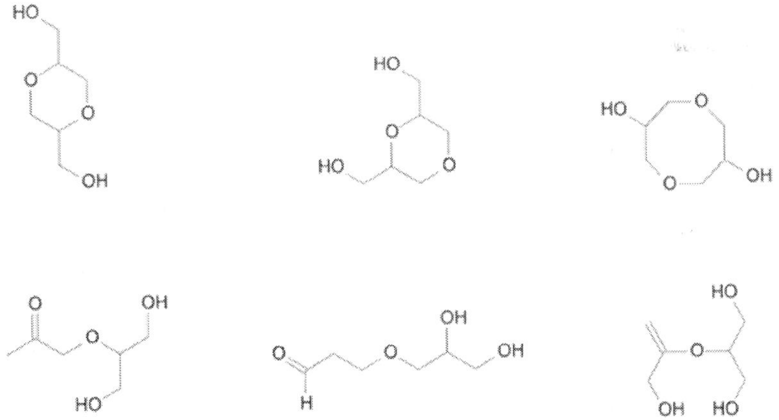

Figure 9. Main diglycerol ethers found in the present work.

The catalyst chosen for reference was sodium hydroxide, which was dissolved in the glycerol, where from the bibliography the amount of NaOH was chosen to give a molar ratio NaOH/Glycerol = 50 [32]. The homogeneous reaction with diluted NaOH (1 g/50 mls glycerol, 240 °C), reached 20 % conversion with glycerol ethers mainly cyclic (18-20 min analysis) after 3 h of reaction, lower temperatures gave very low conversions and higher temperatures or times decreases the selectitivty to diglycerol ethers due to unwanted triglycerol compounds (Figure 10). The reaction with RP45 (1 g/50 mls glycerol) led to a ratio between linear/cyclic diglicerol of 1/4, while on using homogeneous reaction, only cyclic diglyerol was produced under the conditions used (Figure 9). On increasing the NaOH/Glycerol or RP45/glycerol ratios the selectivity decreased due to unwanted glyceric and glycolic acids due to over oxidation and triglycerols [33].

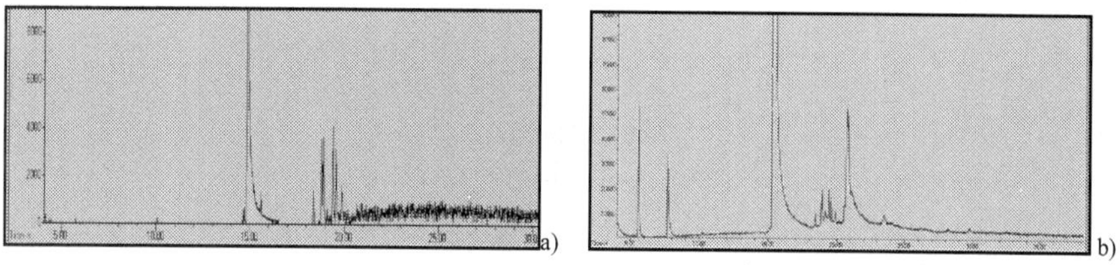

Figure 10. a) Homogeneous reaction of glycerol with NaOH (15 min glycerol, 18-20 minutos cyclic diglycerols). b) Activated RP45 (4 h reaction, 5.7 min: glycolic acid, 7.8 min: glyceric acid, 15 min: glycerol, 18-20 min: cyclic diglycerols, 21 min: linear diglycerols, 23 min: triglycerols).

Given the basicity of RP45 (see Figure 7) it carbontaes easily in the open atmosphere due to reaction with CO_2, therefore it was important to activate this material *in situ* to optimise its activity as a basic catalyst. This was done by heating to 500°C, reaching conversions close to 30 % after 6 h of reaction with selectivity to linear diglycerols close to 50 %. Also, the use of an inert atmosphere was necessary to avoid over oxidation [34]. The reactions carried out with RP47 and RP48 gave very low conversions as corresponds to their smaller active surfaces, since they were prepared at 700 °C and 850 °C respectively (RP47: 8 h, 5 % conversion, 75 % linear diglicerol, 25 % cyclic, RP48: 8 h, 2 % conversion, 50 % cyclic, 50 % linear) [20, 23, 31].

4. Lipases immobilised on materials prepared with agriresidues derived from rice production.

Lipases are enzymes of the hydrolases family, with capabilities such as hydrolysing triglycerides to diglycerides, monoglycerides, fatty acids and glycerol, by reaction of the carboxylic ester bonds. More than 25 % of enzymes used in biotransformations are lipases. However, their high production cost are their main disadvantage for industrial uses like soap and detergent production, baby milk preparation, hydrolysing the grease in milk or production of pharmacological substances. In the human body these substances are important since they facilitate fats adsorption.

Lipase immobilisation is of interest since it allows their reuse and increases their resistance to inactivation. For industrial applications, several properties are important *i.e.* mechanical strength, chemical and physical stability, hydrophobic/hydrophilic character, amount of immobilised enzyme and cost. The use of agroindustrial residues to prepare supports for immobilisation of enzymes can reduce the cost and therefore extend the use of lipases to an industrial scale, since these materaials are cost effective if the technology to make them competitive with commercial ones is developed. In this work the materials used to support enzymes where derived from rice husk (RH) and sunflower (RP) industrial production [13, 35].

The thermal stabilities of the residues in air were analysed by thermal techniques (TGA–DTA) on a Netzsch 409 EP Simultaneous Thermal Analysis device. Approximately 20–30 mg of powdered samples were heated in an air stream of 75 mLmin^{-1} at a heating rate of 5 °Cmin^{-1} from ambient to *ca.* 1000°C, using α-alumina as a reference. The termal data were used to design the controlled thermal treatments to produce agriresidue derived materials at temperatures between 500-700 °C, where the organic matter had been decomposed. The microstructure of the agriresidue derived materials was observed by scanning electron microscopy (SEM) (Hitachi S-4700 type I, Japan). (Figure11).

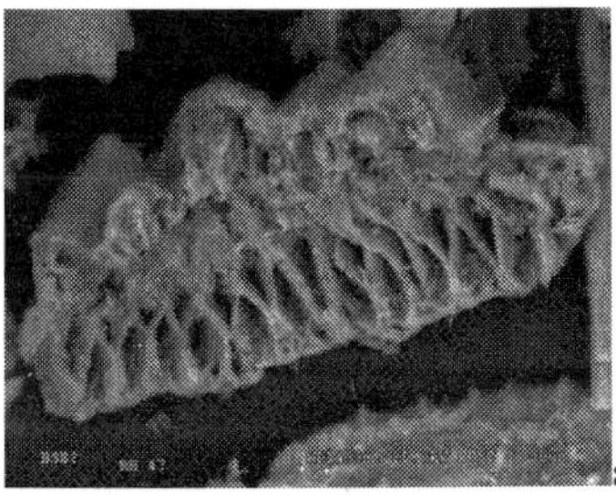
a)

b)

Figure 11. SEM images of rice production derived materials. a) RH47 b) RH26

The TXRF analysis of the materials derived from heat treated rice husk (RH47, RH45, RH26) indicated that they contain *ca.* 39 % silicon and 1-2 % calcium and potassium and those derived from sunflower production (RP45 and RP47) 14 % potassium, 12 % calcium, 7 % magnesium, 2 % phosphorous, 1 % iron and 1 % silicon.

The crystallinity of the materials was recorded by X-ray diffraction (XRD) on a Seifert 3000P diffractometer, using Cu Kα_1 radiation: λ = 0.15406 nm, at 2θ = 5-75 °, with 0.02 ° and 2 sec/pass (Figure 12). According to these analyses, the RH materials have amorphous structures as correspond to their siliceous nature with small amounts of alkaline and alkaline-earth cations. The materials derived from RP residues were crystalline solids with XRD patterns corresponding to oxides of potassium, calcium and magnesium when recently calcined, with increasing cristalinities on heat treatment, and their carbonates when in contact with CO_2 rich atmosphere (fairchildite ($K_2Ca(CO_3)_2$) (red lines), calcite ($CaCO_3$) (blue lines)).

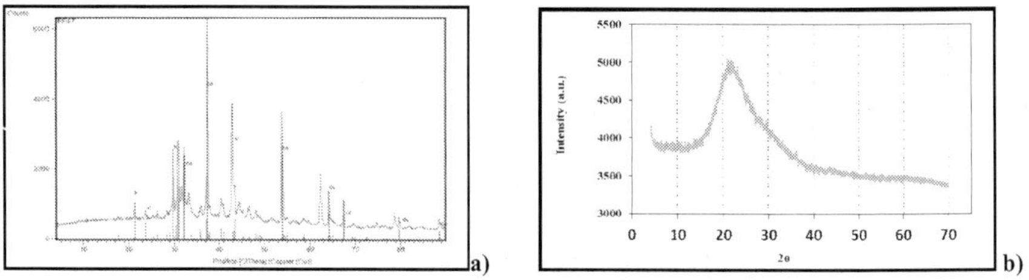

Figure 12. XRD patterns: a) RP47 b) RH26

Fourier transformed infrared transmission spectra (FTIR) of materials obtained on a Nicolet 40 FTIR spectrophotometer in the wavenumber range of 4000–400 cm^{-1}, using a 1/100 dilution in KBr indicated the presence of bands at 900-1200 cm^{-1} and at 400-600 cm^{-1}, corresponding to metal-oxygen bonds (800 cm^{-1} is O-Si-O symmetric stretching vibrations), given the oxide structure of the materials freshly calcined and bands of OH- at *ca.* 2900-3500 cm^{-1}, that decrease on increasing the treatment temperature due to the loss of water from the OH- groups, carbonate groups are found at 1400-1460 cm^{-1} and bands close to 2100 cm^{-1} corresponding to C=O groups present in organic matter, that decrease with the treatment temperature.

The specific surface areas measured by N_2 adsorption at 77 K after outgassing overnight at 150 °C and employing the BET method for data analyses in a Sorptomatic 1800 instrument (Table 4) indicated that RH derived materials are mesoporous with type IV isotherms and wide pore size distributions, and RP derived materails are non-porous with type II isotherms and non-existent hysteresis loops. Specific surface areas are listed below.

Material	SBET m^2g^{-1}
RH26	63
RH45	98
RH47	16
RP25	8
RP47	4

Table 4. Textural analyses by N_2 adsorption desorption and BET calculations

The porosities in pores from 300 μm down to 7.5 nm were determined by mercury intrusion porosimetry (MIP) using CE Instruments Pascal 140/240 porosimeter on samples previously dried overnight at 150 °C, the Washburn equation was employed, assuming a non-intersecting cylindrical pore model and the recommended values for the mercury contact angle and surface tension of 141 ° and 484 mNm^{-1}, respectively. These studies show that materials RH26 and RP47 have pore volumes, mesopores areas, medium pore radii and medium particle sizes as follows: RH26 (0.08 cm^3g^{-1}, 27 m^2g^{-1}, 80 μm and 90 μm), RP47 (0.02 cm^3g^{-1}, 4 m^2g^{-1}, 15 μm and 30 μm).

In order to measure the basicity of the solids acetic acid was previously adsorbed onto the powder materials and subsequently the mas of 44 was recorded against temperature with a quadrupole mass spectrometer, M3 QMS200 Thermostar coupled to Stanton STA model 781 TG/DTA apparatus. For these analyses approximately 50 mg of the materials were dosed with acetic acid, transferred to the crucible placed within the Stanton TG-MS, where they were subsequently flushed with nitrogen gas at room temperature in order to desorb any loosely bound physically adsorbed acetic acid, until a constant weight was attained. The decomposition of the chemisorbed acetic entities was then achieved by increasing the temperature under a nitrogen flow at a heating rate of 5 $°Cmin^{-1}$.

The amount and temperature of evolution of the CO_2 signal gave an indication of the strength and amount of basic sites. The CO_2 signal was calibrated from the decomposition of a known amount of calcium oxalate. These measurements indicated the presence of basic groups of high (>500 °C), medium (300-500 °C) and low basicity (100-300 °C) for RP materials, as corresponds to their content in alkaline and alkaline-earth cations and the RH materials had lower amounts of basic centers and of lower strength than the RP materials (Figure 13) [14].

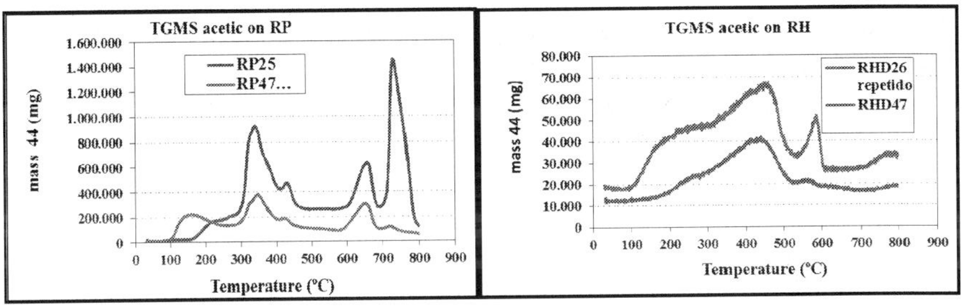

Figure 13. TGMS acetic acid decomposition on RP or RH derived materials.

The immobilisation process and measurement of activity was undertaken until there was no significant variation in activity. Subsequently, the biocatalyst was filtered, washed, dried over P_2O_5 and the enzymatic activity was studied at 30 °C, using a Mettler Toledo (modelo DL-50) pH-stato at pH=8.0 with 0.1 N NaOH titrating agent. As reaction medium 19 mL of 1nM tris-HCl buffer at a pH 8.0, 0.6 mL of acetonitrile and 0.4 mL of tripropionine as reaction substrate. A blank test was done to measure spontaneous hydrolisis (without enzyme and only with the triglyceride and the reaction medum). This technique consists of the controlled addition of a basic solution to maintain the pH, being then the titration proportional to the production of acid and therefore to the reaction rate.

The enzyme immobilisation on RH materials indicated that during the initial hours the percentage of immobilised enzyme grows but after 24 h there was no more absorption, however, when RP materials were used there was a continuous increment of immobilised enzyme, probably due to their lower pore volumes compared to RH materials. The materials that immobilised more enzyme were RH45 and RH26, where the latter was chosen for further studies since it had the highest catalytic activity. For the enzymatic activites, lipase Rhizopus oryzae expressed on levadura Pichia pastori, from the Autonoma University of Barcelona, was used. The lipase received as a solid was used to prepare the enzymatic solution of 20 mg/ml with sodium phosphate buffer 100 mM and pH 6.5, incubated stirring for 1 h at 4 °C and centrifuged to eliminate any solid residue [36].

The measurements of enzymatic activity, in sobrenadantes, control and stock solutions were carried out in a plate reader Versamax, using 10 mM *p*-nitrophenyl propionate (*p*NPP) as reaction substrate, in kinetic mode, with a wavelength of 405 nm, 30 °C and 2 min. Since the data are given in mU/min, the enzymatic activity was calculated with an extinción coeficient (ε) for the *p*NPP appropiate to the wavelength and pH, $\varepsilon = 16780\ M^{-1}cm^{-1}$. The analyses of protein concentration was done by the Bradfor Method using the Biorad reactant and procedure, based on the capacity of dye *Comassie brilliant blue G-250* to change color in the maximum of absorption in the range 465 a 595 nm, according to different concentration of proteins (orange colour becomes blue on the dye bonding to protein at 595 nm. Calibration curves for this procedure were measured with a 50 μg/ml solution of bovine serum albumin (BSA) as standard.

The experiments to study the reactions of hydrolysis were carried out using different enzymes, the test reaction of biodiesel synthesis by transesterification of triglycerides (trilaurin or triolein, Figure 14) with metanol or etanol was done where the main products were ethyl or methyl oleate or laurate and as secondary products mono and diglycerides and the corresponding fatty acid due to triglyceride hydrolysis.

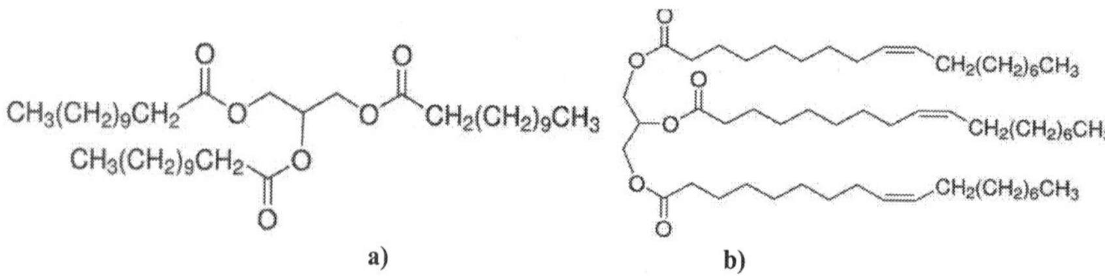

Figure 14. Trilaurin (a) and triolein (b) formulas.

Transesterification reactions were carried out with 50 mM ester concentrations and triglyceride:alcohol molar ratio 1:4, using 2-methtl-2-butanol (2M2B) as solvent, 20 mg/ml of enzyme, 45 °C and 300 rpm stirring speed. The progress of reaction was quantified by means of TLC chromatography, using a solution of hexane, ethyl acetate and glacial acetic acid (90:10:1) and developed by a solution of etanol, water, glacial acetic acid and a dye (*Comassie blue*) (20:80:0.5:0.03) and HPLC composed of a quaternary pump Waters E600, an injector and photodiode detector Varian ProStar and a refractive index detector Waters 2410, with a *Cosmosil* C18 of 4.6 x 150 mm column with a medium particle size of 4.4 μm, at 40 °C with a mobil phase of metanol and water acidified with 0.1 %V acetic acid and variable methanol:water ratio. The analysis method is based on time gradient, varying composition and flow of the mobil phase until only methanol is passed.

Transesterification of triglycerides with methanol and ethanol produce mainly methyl and ethyl oleate and secondary compounds such as fatty acid esters (mono- and dilaurin, monoolein and diolein) and the corresponding fatty acid due to the triglyceride hydrolysis. On using Novozym 435, trilaurin disappears after 24 h of reaction, because when the trilaurin is consumed, ethanol interacts with the diglyceride. Regarding the use of Lipozyme TL-IM, ethyl laureate formation is slower, although trilaurin also gets consumed and the formation of acid is lower, because Lipozyme TL-IM has a lower amount of water than Novozym 435. In the case of Lipozyme RM-IM, the formation of ethyl laureate is not so important as in the cases stated before, and at 24 h there is still a lot of unreacted trilaurin.

Sample	Immobilised Enzyme (%)	Catalytic Activity (U/g)	Recovered Activity (%)	S_{BET} (m^2g^{-1})
RH45	60	393.5	28	98
RH47	46	671.5	60	16
RH26	57	737.2	53	63
RP45	39	308.1	22	8
RP47	49	67.0	4	4

Table 5. Textural and reactivity data for enzyme immobilised on agriresidue derived materials.

Amongst the materials used, those derived from rice husk show higher capacities for lipase absorption. It should be noted that RH45 and RH26 are the materials that have highest surface areas, but taking into account the amount of enzyme absorbed, it is clear that not only the textural characteristics are defining the behaviour of the supports but the amount and strength of the basic sites was also important, much higher in the RP materials. RH26 was chosen for preparation of immobilised enzyme for biodiesel synthesis in comparison with commercial enzymes. The ROL enzyme supported on RH26 prepared in the group had a higher activity than Novozyme 435 and similar to that of Lipozyme TL-IM, both lipases widely used in biocatalysis. Trilaurin was consumed in *ca.* 24 h with ethyl laureate formation close to that of Novozyme 435 and lauric acid due to the hydrolysis produced by the water contained in the enzyme.

	RH26	Novozyme 435	Lipozyme TL-IM	Lipozyme RM-IM
U/g	1347.2	1321.5	1444.5	95.4

Table 6. Comparison of reactivity of commercial and agriresidue supported enzymes.

The biocatalysts prepared by immobilisation of the lipase on agriresidue derived materials, given their renewable origin and low cost, seem to be an attractive option for reducing costs and environmental impact of these processes [37, 38].

5. Bioecomaterials

5.1. Tissue engineering (Dental and Bone replacement therapies)

Bone tissue engineering is one of the most promising approaches to be used as an alternative to conventional autogenic or allogenic surgical techniques for bone tissue repair [39]. Bone grafts are used to stimulate the formation of new bone in many conditions such as congenital anomalies, cancers, and trauma or to improve the regeneration of bone tissue around surgically implanted devices.

An ideal bone graft or scaffold should be made of biomaterials that emulate the structure and properties of natural bone extracellular matrix providing all the necessary environmental cues present in natural bone. The tissue regeneration capacity of these bone grafts is measured in terms of their osteogenic, osteoconductive and osteoinductive potential. The osteogenic potential of a bone graft is given by cells involved in bone formation, such as mesenchymal stem cells, osteoblasts, and osteocytes. The term osteoconductive refers to the scaffold or matrix which stimulates bone cells to grow on its surface. Osteoinductive capacity of a bone graft is perhaps the most important property in bone healing as it refers to the stimulation of mesenchymal stem cells to differentiate into preosteoblasts to begin the bone-forming process [40].

A possible therapy for the treatment of skeletal defects has arisen with the use of synthetic materials as bone substitutes. Tissue engineering strategies based on the use of biocompatible and biodegradable porous materials that act as structural templates or scaffolds to guide the growth and development of new bone tissue, supporting both extracellular matrix formation and cell-cell interactions [41]. Due to their similarity to the chemical composition of bone, calcium phosphates can be used to regenerate osteoporotic bone as coatings that improve orthopedic implants, and for odontostomatologic applications, where their particle and crystal sizes are important parameters to be controlled to optimise these processes [42]. Calcium phosphate-based scaffolds exhibit osteoconductivity, bioactivity and resorbability *in vivo* due to their complex chemical composition (Ca/P ratio) and physical properties such as crystallographic structure and porosity [40]. However, major drawbacks in the use of synthetic calcium phosphates are their price and use of non-renewable resources.

Biomaterials based on composites of calcium phosphates and silica have the ability to bond directly to bone and thus enhance bone formation through supply of chemicals needed to support cell function and tissue formation. Furthermore, it has been found that addition of silica to the calcium phosphate scaffolds was beneficial in increasing the mechanical strength, cellular proliferation and dissolution/resorption rates [43, 44]. Moreover, the presence of magnesium in these materials favours bone growth, promoting osteogenic differentiation of preosteoblasts and improving osteointegration during the early stages of bone healing. During the *in vivo* degradation of the scaffolds cell proliferation and differentiation are promoted by the release of their component elements [45-47].

With the aim to convert waste into value-added products agricultural wastes (such as beer bagasse) have been investigated as potential renewable raw materials to develop bone scaffolds capable to support osteoblast growth for bone regeneration applications. Materials prepared here with residues from beer production contain P, Si, Ca and Mg as main components, which are also cations present in bone [11, 48]. Furthermore, the use of agricultural wastes to provide renewable raw materials for more advanced applications is of great interest giving value-added products that may lead to a significant reduction in waste accumulation. Moreover, due to their origin these materials are very competitive in price [49]. The materials derived from beer bagasse (BBM) were biphasic calcium-magnesium phosphates with silica that can be either amorphous or as cristobalite, its crystallinity increasing in accordance with the final heat treatment temperature employed, with porosities that lie within the 10 to 100 μm range. All of these characteristics were important for the promotion of both cell proliferation and differentiation [50].

Osteogenic cells MC3T3-E1 are widely employed to study *in vitro* matrix mineralisation, since these cells can differentiate into osteoblasts that express strong ALP activity and can form a collagenous matrix organised in 3-dimensional nodules, which in the presence of ascorbic acid and phosphate progressively become mineralised [51]. The MC3T3-E1 cells display a time-dependent sequential expression of osteoblast characteristics that are analogous to *in vivo* bone formation [52]. It has been shown that some biomaterials are able to modify directly the osteoblastic proliferation rate and its differentiation, such as the synthesis of alkaline phosphatase, matrix mineralisation and collagen secretion [53,54]. Thus, the *in vitro* proliferation and differentiation responses of this osteoblast like cell line (MC3T3-E1) to the BBM derived powders were studied. Several biological responses to the biomaterials were assayed, including determinations of cell viability by the MTT and LDH assays, evaluation of ALP activity, Type- I collagen secretion and evaluation of matrix mineralisation at the differentiation period.

The present work employs residues from beer production from three different Spanish plants, Lerida, Guadalajara and Burgos from the Mahou San Miguel group. These residues were chosen so that a comparison of their suitability and any effects of the differences in their chemical compositions on their cytocompatibility for bone growth could be determined. The beer bagasses were first dried at 150°C for 4 h, at a heating rate of 5 °C/min, in order to avoid putrefaction, due to their high humidity (70–85 wt%). Thermal stabilities of the dried materials were determined by TG-DTA analyses in air, to assess the temperature necessary to eliminate the organic matter and prepare stable and reproducible materials. Results from TG-DTA indicated that the thermal behaviour of the three samples under air treatment was practically identical, with total weight losses found of *ca.* 97 % of the initial mass, 8 % due to loss of volatile matter and water at $T < 200$ °C, 55 % loss at T = 200-380 °C caused by the decomposition of organic matter (mainly cellulose and hemicellulose) and finally a 34 % lost for T = 380-600 °C, corresponding to the decomposition of lignin [48,49]. For a more detailed explanation the results obtained with sample BBM Lerida are shown in Figure 15b.

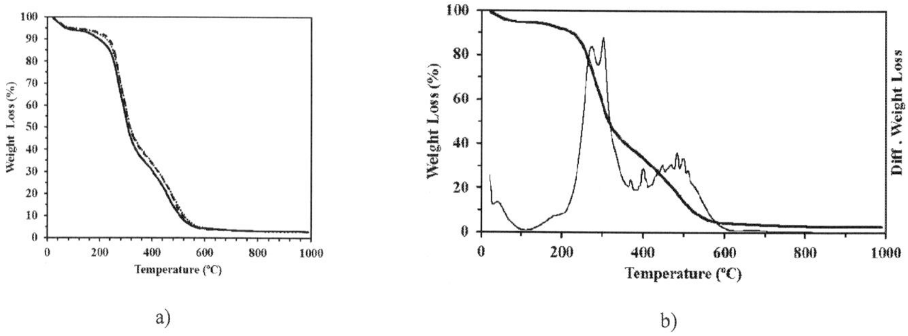

Figure 15. Reproducibility of beer bagasse from a) Lerida, Guadalajara and Burgos and detailed analysis of b) BBM-Lerida.

The samples of bagasse were then calcined at 600, 700, 850 or 1000 °C, maintaining the final temperature for 4 h, the samples thus produced were designated as BBM46, BBM47, BBM48 and BBM410, respectively. The calcined materials were homogenised and their particle sizes controlled by being milled to less than 120 μm, due to the importance of this parameter in the reproducibility of biological behaviour of the materials.

The composition of the calcined materials was analysed by means of inductively coupled plasma atomic emission spectroscopy (ICP), showing four main elements that depending on the source of the beer bagasse were 15–20 % silicon, 12–14 % phosphorous, 7–8 % calcium and 5–7 % magnesium. The variations between different batches of beer bagasse from the same source were negligible and within experimental error. X-ray diffraction (XRD) patterns of samples showed that with higher heat-treatment temperatures the XRD peaks were narrower and better defined due to the increased crystallinity of the materials. The most significant crystalline phases were calcium-magnesium phosphate (*) (31.5 ° (100 %), 29.7 ° (85 %) and 29.3 ° (75 %) present at all temperatures and cristobalite which was only found when heat treatment temperatures greater than 600 °C were employed (21.9 ° (100 %, (111)), 36 ° (12 %, (220)), 31.3 ° (10 %, (102)) and 28.4 ° (8 %, (111)) [50, 51].

No crystalline cristobalite was found for BBM46 but mean crystallite sizes of 56 to 70 nm, 60 to 85 nm and 85 to 230 nm were found for BBM47, BBM48 and BBM410, respectively (Figure 16).

The three beer bagasses as received had identical FTIR traces and the heat treated materials prepared from them also displayed identical results. For the bagasse dried at 150 °C the principal bands were due to a broad band of O-H stretching in the 3100-3600 cm^{-1} region, the C-H aliphatic axial deformation in CH_2 and CH_3 groups from cellulose, hemicellulose and lignin at 2926 cm^{-1}, the $-OCH_3$ vibration at 2854 cm^{-1} due to lignin or hemicelluloses. The C=O stretching of the acetyl groups present in cellulosic material at 1743 cm^{-1}, while the bands at 1043 and 1160, corresponded to O-H stretching of primary and secondary alcohols, respec-

http://dx.doi.org/10.5772/60018

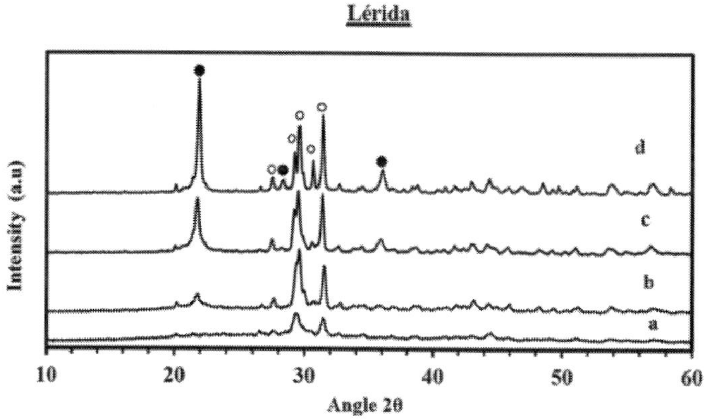

Figure 16. XRD patterns of Lerida agriresidue derived materials heat treated at 600 °C a, 700 °C b, 850 °C c and 1000 °C d.

tively and the band at 1378 cm^{-1} corresponded to O-H vibration of phenolic groups. The signal at 899 cm^{-1} was assigned to β-glycosidic linkages between monosaccharide units. For the heat treated materials the broad band at 3466 cm^{-1} was due to the stretching vibration of the P-OH and Si-OH groups that was diminished with respect to the dried materials due to the loss of these groups on heating. The bands at 1164, 1121 and 1097 cm^{-1} were due to the Si-O-Si asymmetric stretching vibration, a band at 470 cm^{-1} associated with a network O-Si-O bond bending modes. The band at 1023 cm^{-1} was attributed to the symmetric terminal P-O stretching mode of the calcium magnesium phosphate, a band for asymmetric bridge P-O stretching mode appeared at 964 cm^{-1} whilst those at 579 and 496 were due to the asymmetric bending vibrations of terminal P-O bands [52-54].

The porosities and particle size distributions of the heat treated materials were determined by Mercury Intrusion Porosimetry. From Figure 17 it may be appreciated that the majority of the intrusion curve was due to interparticulate pore filling and that only with the materials treated at the lowest temperature was there any sign of mesoporosity, pores with diameters lower than 50 nm, due to intraparticulate porosity, which due to sintering of the materials disappeared on heating at higher temperatures.

As the heat treatment temperature was raised the density of the materials increased with a corresponding reduction in the cumulative pore volume accompanied by a slight displacement of the curves to wider pores. For finely divided powder samples the cumulative intrusion curve represents the void filling between the aggregates of the primary particles. Thus, an evaluation of their size may be made using the Mayer Stowe theory that relates the porosity of the sample to a packing factor, assuming spherical particle geometry, which is used to estimate the particle size from the measured width of the spaces between the particles. It may be observed from the results that higher heat treatment temperatures caused an increase in the aggregate sizes and a densification of the materials due to sintering of the samples, which was in agreement with the results observed from the XRD analyses of these materials [55,56].

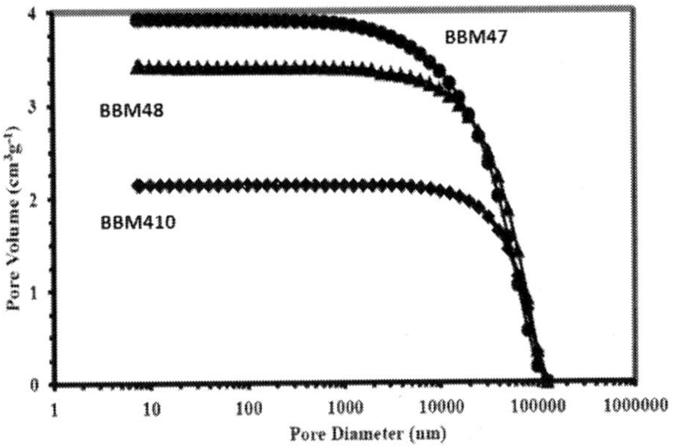

Figure 17. Mercury intrusion porosimetry results for beer bagasse from Burgos treated for 4 h at 700 °C, 850 °C or 1000 °C.

Analyses of the basic character of materials by decomposition of acetic acid (Figure 18) indicate higher amount of basic sites for those prepared at lower temperatures, agreeing with the sintering process observed by the other characterisation techniques.

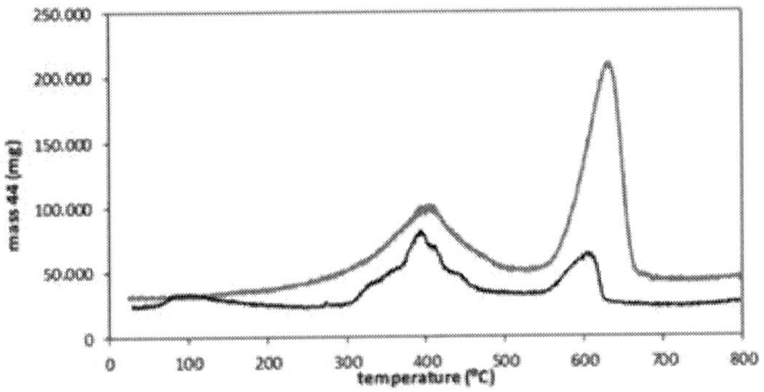

Figure 18. TGMS analyses of decomposition of acetic acid on BBM47 (grey line) and BBM410 (black line).

Cell proliferation and differentiation on BBM derived materials

Cell cultures

The osteoblast-like MC3T3-E1 murine cells were cultured in α-MEM (Gibco) that was supplemented with 10 % foetal bovine serum and 1 % penicillin-streptomycin (basal medium). In order to induce differentiation the cells were placed in osteogenic media: a basal medium supplemented with 10 mM β-glycerophosphate and 50 μg/mL ascorbic acid. These cells were incubated at 37 °C in a humidified atmosphere and at 5 % CO_2. The cell culture results obtained were compared with a reference material, hydroxyapatite (HA), a synthetic calcium phosphate ceramic that mimics the natural apatite composition of bones and teeth and has been described as a potential material to coat scaffolds for promoting osteoblast differentiation [55,56].

Cell proliferation on BBM derived materials

Cell proliferation assays were performed in the presence of increasing concentrations of BBM derived materials from Lerida, Guadalajara and Burgos treated at increasing temperatures, after culturing cells in basal medium for 7 days, in order to determine the influence of the origin of the bagasses and the effect of the temperature to which they were subjected. To evaluate the proliferation rate of MC3T3-E1 cells grown in the presence of BBM derived materials, the cell viability was measured following incubation of the cells with materials at various concentrations, ranging from 20-200 μg/mL for 7 days. HA was used at the same concentrations as a reference material. To carry out the viability assays the cells were seeded into 96-well plates (10000 cells per well; four replicates for each condition). After 24 h, the cells were treated with materials for the specified concentrations and time periods. The cultures were then washed twice in phosphate buffered saline to remove any residual material. Subsequently, the tetrazolium dye, 3-(4,5-dimethilthiazol-2)-2,5-diphenyl-2H tetrazolium bromide (MTT, 5 mg/mL in phosphate-buffered saline; Sigma), was added to the medium and left for 1 h. Following removal of the medium, the precipitated formazan crystals were dissolved in optical grade dimethyl sulphoxide (200 μL). Then by use of an ELX808 microplate reader (BioTeK) the absorbance of each well was measured spectrophotometrically at 570 nm. When beer bagasses were treated at lower temperatures a decrease in cell proliferation rates was observed (Figure 19).

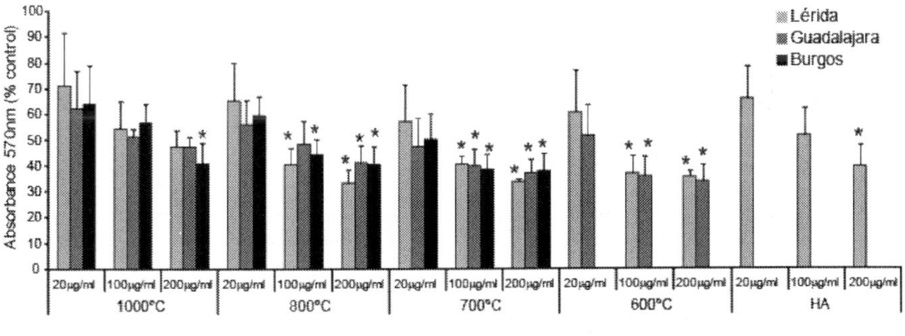

Figure 19. MC3T3-E1 proliferation analysis by MTT assay on cells treated with different concentrations of BBM derived materials treated at several temperatures after one week in culture. The data were analysed by single factor analysis of variance followed by the *post hoc* Tukey's honestly significant difference test, *$p<0.05$ with respect to control (cells growing on polystyrene plates).

We found that the addition of BBM derived materials to MC3T3-E1 cells did not considerably alter the viability of the cells, compared to the reference material (HA). When BBM derived materials were subjected to 1000 °C for 4 h, proliferation analysis assessed by MTT test showed a similar cell growth to that obtained when using HA as reference material. However, a decrease in cell proliferation rates was observed when BBM derived materials were treated at low temperatures (600 °C). The observed delays in cell proliferation were due to the more basic pH for the powder containing media at the first day in culture, which could initially result in a lower cellular enzymatic efficiency and hence in slower processes (e.g. cell division and metabolism) than that observed with the control cells, growing on polystyrene plates [57]. The different relative metabolic levels found in MC3T3-E1 cells growing in the presence of BBM derived materials at day 7 also correlate with the characteristics of the powders, where materials pretreated at lower temperatures produced higher increases in the pH of the culture medium due to their greater basicities, leading to lower proliferation rates of the MC3T3-E1 cells.

According to these findings, further cytotoxicity assays and cell differentiation experiments were performed on the BBM derived biomaterials treated at 1000 °C.

Cytotoxicity

The cytotoxicity of the culture media was related to the lactate dehydrogenase (LDH) activity. The measurements were determined on cells plated at a density of 10 000 cells per well in 96-well plates in basal and osteogenic medium. Beer bagasse materials treated at 1000 °C for 4 h were added at 100 μg/mL. After 24 h, the culture media were collected and centrifuged and the supernatant was used for the LDH activity assay. The LDH activity was determined spectrophotometrically using the Cytotoxicity Detection kit (Roche), according to the manufacturer's instructions. Cells cultured in the presence of BB-derived materials showed no obvious cytotoxicity compared to cells grown in the presence of hydroxyapatite and polystyrene culture plates, used as controls (Figure 20).

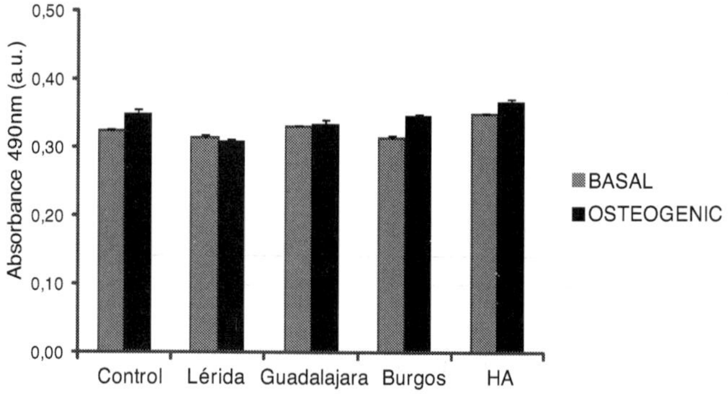

Figure 20. Cytotoxicity study by lactate dehydrogenase (LDH) activity assay on MC3T3-E1 cells treated with 100 μg/ml of BBM derived materials treated at 1000 °C and HA, after 24 h in culture.

Results obtained in LDH assay when MC3T3-E1 cells were cultured in the presence of BBM derived materials for 24 h support the cytocompatibility of these new materials.

http://dx.doi.org/10.5772/60018

Cell differentiation

Previous studies have demonstrated that expression of osteoblastic markers in MC3T3-E1 cells begins after culturing the cells with medium supplemented with β-glycerol-phosphate and ascorbic acid [52]. The effects of direct contact of BBM derived materials and osteoblast-like cells in terms of cell differentiation were evaluated by testing alkaline phosphatase (ALP) activity, collagen production and extracellular matrix mineralisation after 15 days in culture. Alkaline phosphatase activity (ALP) begins to be expressed after 1 week and reaches a maximum after 2 weeks when MC3T3-E1 cells are cultured in osteogenic medium [52].

The capacity of cells growing in the presence of BB-derived materials was evaluated to express alkaline phosphatase, an early marker of osteoblastic cell differentiation. To this end, MC3T3-E1 cells were seeded into 96-well plates (10 000 cells per well; four replicates for each condition) and grown on basal and osteogenic medium for 15 days in the presence of BB-derived materials at 100 µg/ml. After treatment with the BBM derived materials, the cells were rinsed with PBS and then lysed into PBS containing 0.1 % Triton X-100. These cell lysates were then centrifuged and the soluble fraction used for the enzyme assay. The samples were first incubated with an assay mixture of *p*-nitrophenyl phosphate (p-NPP) (Sigma). Cleavage of the *p*-NPP in a soluble yellow end product, *p*-nitrophenol, which absorbs at 405 nm, was used to assess the ALP activity. The optical density of *p*-nitrophenol at 405 nm was then determined spectrophotometrically and the ALP activities normalised to total protein content using the bicin-choninic acid (BCA) method. The ALP activity of each condition was quantified and compared to that present in cells grown on polystyrene plates used as the (control).

It was found that the ALP activity was higher in cells grown in osteogenic medium than for cells cultured in basal medium, as we would predict (Figure 21). However, there was no significant difference in the ALP activity observed for MC3T3-E1 cells grown in the presence of HA and with control cells. These results established that the presence of BB-derived materials did not affect ALP activity.

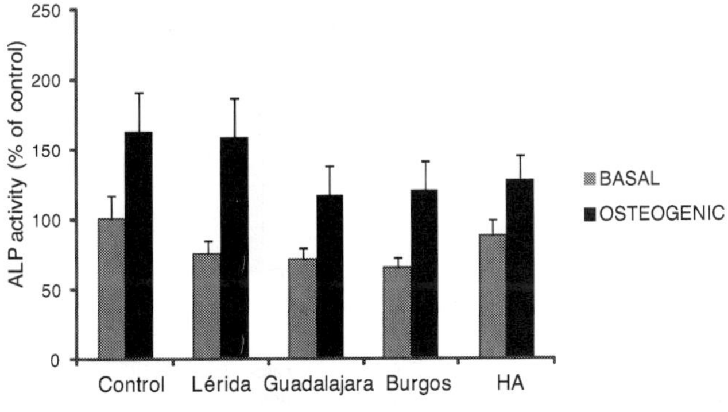

Figure 21. MC3T3-E1 differentiation study by alkaline phosphatase (ALP) activity assay on cells treated with 100 µg/ml of BB-derived materials and HA after 2 weeks in culture.

Results indicated that MC3T3-E1 cells maintain their capability to express active ALP enzymes when growing in the presence of BBM derived powders.

The addition of ascorbic acid in MC3T3-E1 cells is known to induce the deposition of collagen in the extracellular matrix [52]. To confirm that osteoblastic cells exposed to the BB-derived materials indeed maintained the ability to differentiate at similar levels to control cells, the profile of type-I collagen cellular secretion, the main extracellular matrix protein expressed in bone, was also analysed. Collagen secretion by MC3T3-E1 cells cultured in the presence of BB-derived materials treated at 1000 °C for 4 h, at 100 μg/mL was quantified by Sirius Red staining. After culturing MC3T3-E1 cells in the presence of BB-derived materials for 15 days in both basal and osteogenic media, the cells were washed three times with PBS and then fixed in 4 % paraformaldehyde. Following the three rinses in PBS, the cell cultures were stained for collagen secretion in a 0.1 % solution of Sirius Red (Sigma) in saturated picric acid for 18 h. Following washing with 0.1 M acetic acid until the disappearance of the red colour, the stain on specimens was eluted in destain solution (0.2 M NaOH–methanol 1:1). The optical density at 540 nm was then determined using a spectrophotometer.

From the results shown in Figure 22, it may be seen that collagen deposition was promoted when MC3T3-E1 cells were grown in osteogenic medium at all the tested conditions, as expected and no significant differences in the collagen production were observed for cells grown in the presence of BB-derived materials compared with those grown on plastic plates, neither in basal nor osteogenic medium.

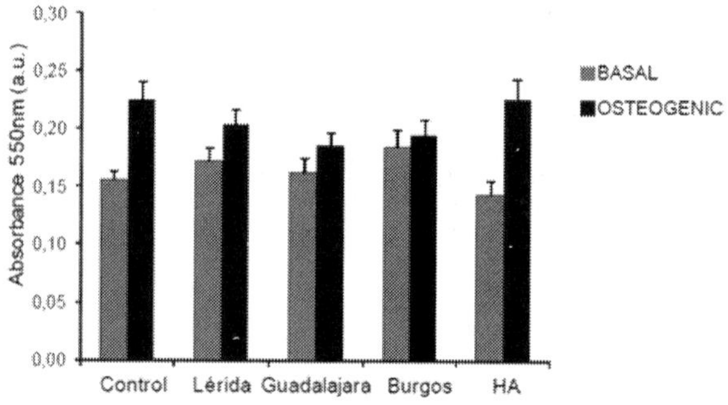

Figure 22. MC3T3-E1 differentiation study by collagen production (Sirius Red staining) on cells treated with 100 μg/ml of BB-derived materials and HA after 2 weeks culture.

Extracellular matrix mineralisation is also one of the major aspects of bone formation. Minerals formed *in vitro* were found to consist of calcium and phosphorus deposited on well-bonded collagen fibrils, and some of the crystals matured into hydroxyapatite crystals [58]. Besides its effect on ALP activity and collagen synthesis, we determined whether BB-derived materials might affect the mineralisation of the matrix formed by MC3T3-E1 cells. These cells are known to deposit minerals in the collagenous matrix in the presence of β-glycerol phosphate [59].

Extracellular matrix calcium deposits for mineralised nodule formation were stained with Alizarin red S dye which combines with calcium ions. After culturing MC3T3-E1 cells in the presence of BBM derived materials at 100 μg/mL for 15 days, the cells were then washed thrice with PBS and subsequently fixed in 75 % ethanol for 1 h. These cell cultures were then stained with 40 mM Alizarin Red S in distilled water (pH 4.2) for 10 min at room temperature. The cell monolayers were then washed with distilled water until no more colour appeared. The stain was dissolved in 10 % cetylpyridinium chloride in 10 mM sodium phosphate (pH 7.0) and the absorbance values at 620 nm were measured. The extracellular matrix mineralisation determined by Alizarin Red S staining is shown in Figure 23. In all of the tested conditions cells grown in osteogenic medium for 15 days displayed slightly higher calcium content, an indicator of mineralisation nodule formation. Treatment with BB-derived materials at a concentration of 100 μg/mL for 15 days did not significantly affect the mineralisation rates compared with those of control cells grown on polystyrene plates (Figure 23a). For these determinations HA (a material which contains Ca/P-apatite) could not be used as a control due to the high background produced by the material itself. These results indicated normal mineralisation induced by MC3T3-E1 cells in long-term cultures was not affected by the BB-derived materials.

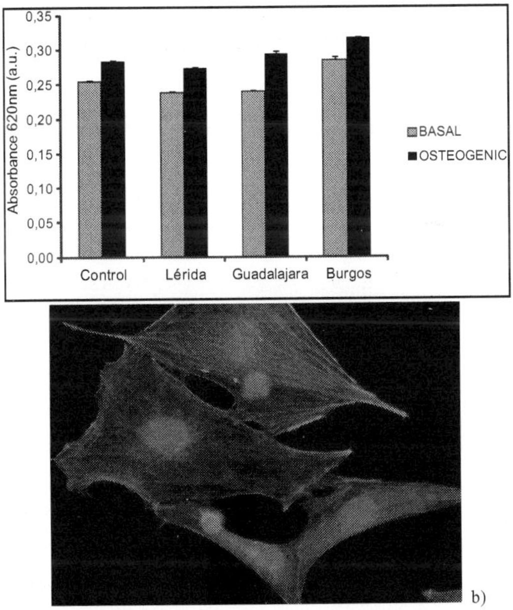

a)

b)

Figure 23. a) MC3T3-E1 differentiation study by extracellular matrix mineralisation (Alizarin Red S staining) on cells treated with 100 μg/ml of BBM derived materials after 2 weeks in culture. b) Osteoblasts (MC3T3-E1cells) growing in the presence of BBM derived materials. Cells were stained with phalloidine (red) and the nuclei were counterstained with ToPro-3 (blue).

Overall, in the presence of BB derived materials the osteoblast functions displayed normal cell differentiation profiles with respect to alkaline phosphatase activity, collagen secretion and extracellular matrix mineralisation. Furthermore, these parameters were maintained when compared to control cells grown on plastic plates and also with those obtained by culturing MC3T3-E1 cells in the presence of the same amounts of HA. The use of these BB-derived materials as coatings of metallic and ceramic bioimplants for odontoestomatologic treatments, or to form part of 3D scaffolds with pore sizes designed with the desired characteristics for bone replacement, is currently under study. Together the results suggest that the developed materials may potentially be used to prepare scaffolds for bone tissue engineering applications.

5.2. Controlled desorption of bioactive substances

The biological activity of a substance depends mainly on the nature of its interaction with the tissue or organ; it must reach the target in an amount that is adequate to produce the desired effect, which means that it should be liberated in the particular place at a controlled rate. The same amount of active ingredient can have different effects when it is formulated as oral solution or in capsules or pills, due to the different rates of adsorption of the active agents in the digestive track. These requirements mean that preparing drugs from pure substances is a multidisciplinary and extensive field that requires multidisciplinary expertise in pharmaceutical sciences, engineering, material sciences, physical chemistry, polymer science, solution chemistry and biochemistry, amongst others.

The work presented here is based on the use of beer and rice production residues to prepare materials with special characteristics towards controlled desorption of bioactive substances, using the anticarnogenic drug 5-Fluorouracil as the model molecule (Figure 24).

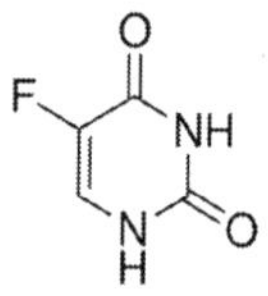

Figure 24. 5-Fluorouracil, anticarcinogen

The drug 5-fluorouracil (5-FU) is a pyrimidine analogue, used in the treatment of cancer, as it is an irreversible inhibitor of thymidylate synthase; interrupting the action of this enzyme blocks synthesis of the pyrimidine thymidine, nucleoside required for DNA replication [60]. 5-FU belongs to the World Health Organization's List of Essential Medicines, being part of the family of drugs called antimetabolites. It has been used amongst others in the treatment of breast, stomach, pancreatic and skin cancers. Its main disadvantage is that the same dose of 5-FU may have therapeutic response with low toxicity in some patients, while even life-threatening toxicity in others [61]. Thus, its use in a controlled manner is of great interest to avoid these problems. Parenteral administration causes a rapid elimination of 5-FU with an apparent terminal half-life of approximately 8-20 min. Choosing a proper controlled release system can improve its anticancer activity and also decrease the adverse side effects. 5-FU is a neutral weak acid [62, 63] whose tautomers structures are shown in Figure 25.

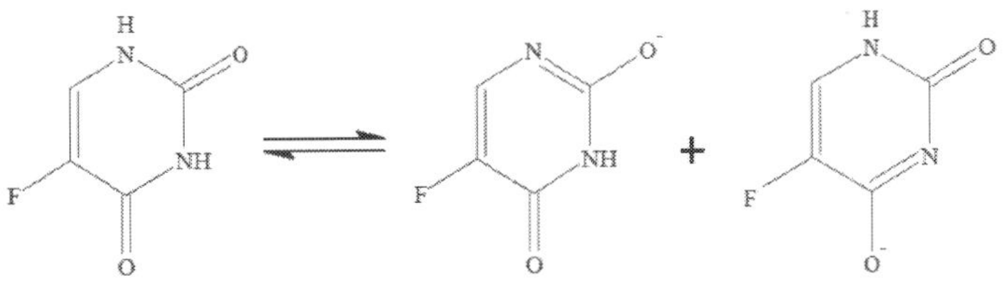

Figure 25. 5-FU tautomers

The design of materials to be used in desorption of bioactive substances is based on a thorough study of the conditions necessary to achieve a texture and structure capable to induce desorption in a controlled manner. The residues were first treated thermally to avoid putrefaction, as indicated above and then, according to TG-DTA analyses three different temperatures were chosen, *i.e.* 700, 850 and 1000°C, after previous studies by infrared spectroscopy (FTIR), textural analysis, X-ray diffraction and acetic acid decomposition on basic sites by TG-MS, and textural analyses by N_2 adsorption/desorption and mercury intrusion porosimetry. In this way the residue derived materials have different structure and surface characteristics, with higher crystallinities and lower surface areas, porosities and basic sites on their surfaces, on increasing the treatment temperature.

The biocompatibility of these bioecomaterials BBM47, BBM48 or BBM410 was studied after crushing and sieving, and homogeneized to a controlled particle size to favour reproducible results, according to previous studies. In order to assess the bioecomaterials biocompatibility a human glioblastoma cell line (1321N1) was used (Figure 26).

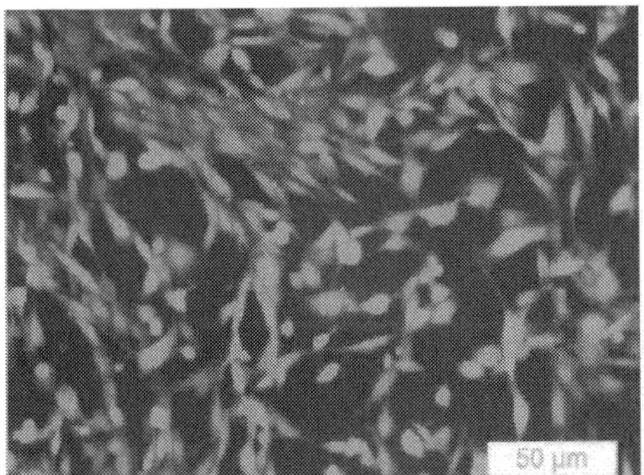

Figure 26. Images of 1321N1 cells stained with AM-calcein showing live cells.

The viability of 1321N1 cells growing in the presence of the materials was determined at 1, 3 and 7 days.The quantification of cell viability on the different materials (Figure 27) indicates that all the materials were biocompatible, with cell viabilities similar or even better than the plastic control up to 7 days.

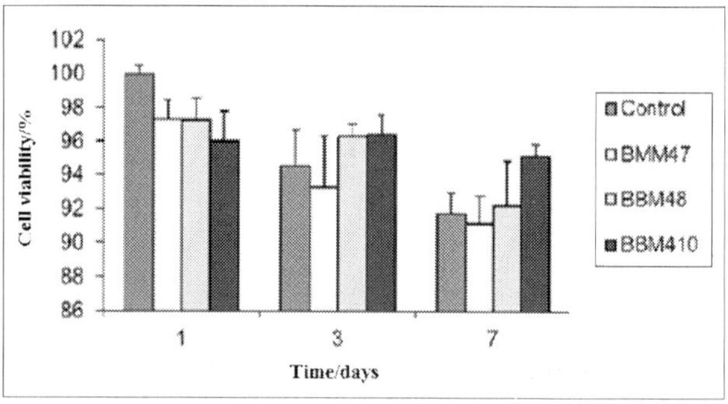

Figure 27. Cell viability (%) determined by MTT assay showing cells growing on biomaterials after 1, 3 and 7 days.

The preparation of the bioecomaterials to be used in controlled desorption was undertaken by studying different important parameters in their structuration, such as pelletising conditions, use of porogens, *etc*.. The particle size was controlled by sieving, and the conditions of pelletising (1 cm diameter) studied by changing the amount of material, pressure and time, being optimised at 5 tons pressure for 2 min. The material used was BBM47, since this had a higher surface area and porosity leading to a greater capacity for 5-FU absorption.

After the pellets were prepared, they were sintered, once again after the corresponding study of the temperature and time of sintering, since they are parameters of utmost importance. Experimental results indicated that the optimum temperature of sintering was 700 °C for 4 h. Lower temperatures or times did not produce pellets with enough cohesion and higher temperautres or times produced materials with lower surface areas and pore volumes.

To determine the 5-FU concentration needed to eliminate 1321N1 cells after 2 days in cultures, the MTT assay was used, as described above. The 1321N1 cells were cultured (20000 cells/well in 24-well plates) at 37 °C in 5 % CO2 with 95 % humidity atmosphere for 24 h to favour their adhesion to the plates. Afterwards different concentrations of 5-FU were added to the medium and 2 days later the MTT assay was carried out.

From the results shown in Figure 28, it can be observed that a 5-FU concentration of 10 μg/mL was enough to eliminate all 1321N1 cells present in the culture. However, a concentration of 20 μg/mL was chosen, since it was possible that not all the adsorbed 5-FU would be liberated. Consequently, 200 μL of a solution 5-FU (20μg/mL) were added to each pellet, left to dry at room temperature for 24 h, protected from light, 5-FU is light sensitive. Finally the pellets were left in contact with human glioblastoma cells 1321N1, growing in DMEM (Gibco) supple-

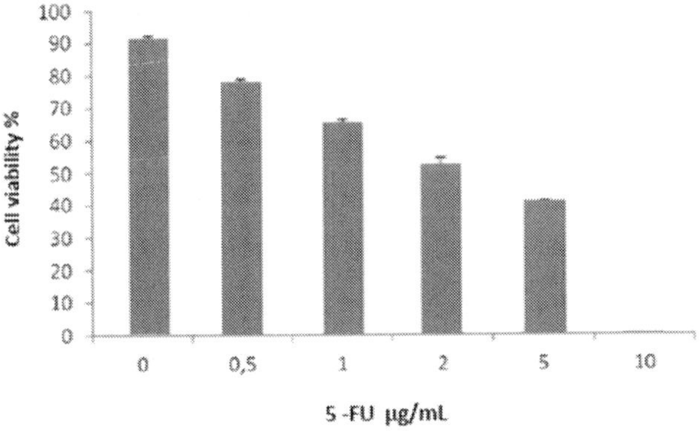

Figure 28. Cell viability after 2 days of treatment with different amounts of 5-FU (µg/mL).

mented with 10 % of bovine phoetal serum and 1 % peniciline/estreptomicine and the cell viability studied with fluorescent probes, as indicated above, following the decrease in cell viability as the 5-FU was desorbed from the pellets. The analyses of the desorption kinetics was carried out by comparison with the data obtained using a mesoporous silica prepared in our research group from rice husk (MR). This mesoporous material contains more than 97% silica, which is considered as ideal for desorption of pharmaceutical compounds [64].

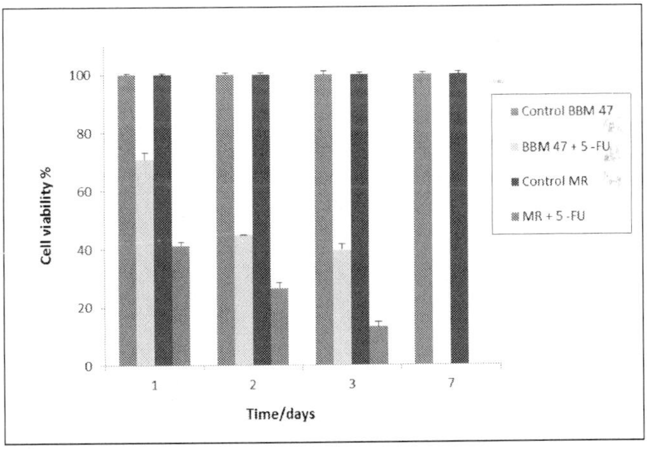

Figure 29. Viability of 1321N1 cells after 1, 2, 3 or 7 days in contact with BBM47 bioecomaterial or MR with 5-FU adsorbed.

According to these results, it can be said that 1321N1 cells could be eliminated with both materials, although BBM47 eliminates cancerous cells in a more controlled manner than MR. The characterisation of these materials indicates that MR was an amorphous mesoporous silica with a surface area of 98 m^2g^{-1} and BBM was crystalline with silica in the form of cristobalite and calcium and magnesium phosphate of 4 m^2g^{-1}, also the basicity of BBM was much higher than that of MR, given its composition rich in alkaline earth cations (Figures 13 and 18), versus MR having more than 97% silica. The importance of the basic properties of the materials for the controlled release of 5-FU was to be expected, since given the acidity of 5-FU it would adsorb on basic sites. This explains why even though BBM47 had a much lower surface area than MR, its much higher basicity leads to a greater interaction with 5-FU, facilitating its controlled desorption, after careful design of its textural and structural characteristics [62, 65-67].

6. Conclusions

Residues from the beer, rice and sunflower oil production industries were used to prepare materials capable of cleaning wastewaters, support enzymes, act as catalysts or as scaffolds for tissue engineering or controlled desorption of bioactive substances. The use of these waste materials gives them an added value as a sustainable, environmentally friendly and economic supply of nanostructured materials, *i.e.* BBM derived materials obtained were as cytocompatible and osteogenic as HA powder, commonly used in bone and teeth replacement therapies, being non-cytotoxic and supporting cell growth and differentiation and of controlled desorption of anticarcinogen 5-Fluorouracil. Sunflower oil production residues can be used as raw materials to transform glycerol from biodiesel production of the same company towards fuel additives. Rice husk derived materials can be used to support enzymes in an inexpensive and environmentally sound way. Furthermore, the design of materials are based on preparation methods kept as simple and inexpensive as possible, with low energy consumption and easily adoptable procedures, by careful design of parameters and use of characterisation techniques used in materials engineering.

Acknowledgements

The authors wish to acknowledge the financial support from the CDTI projects "Systems to recycle aqueous effluents with valorisation of contaminants and reduction of discharges. Total oxidation of contaminants in effluents by using Ecomaterials" (CDTI IDI-20091139), "Development of applications of glycerine sub product of biodiesel with catalysts made from Ecomaterials" (CDTI IDI-20121298) and "Design of sílica with controlled texture to be used in detergency using as raw material rice husk ash" (CDTI IDI-20143711-20143713) and the Spanish National Science Council for the projects "Valorisation of agriresidues for generation of biocompatible materials (Ecomaterials) to be used in dental implants" (INNPACT IPT-2011-1935-310000), "Preparation of materials from agro industrial residues for biomedical uses" (PIE CSIC 201480E103) and "Valorisation of Spanish agriresidues for advanced uses" (PIE CSIC 201460E105).

http://dx.doi.org/10.5772/60018

Author details

A.M. Martínez Serrano[1,2,3], M. Ramos[1], M. Yates[2], M.A. Martin-Luengo[3*], F. Plou[2], J.L. Lacomba[4], G. Reilly[5], C. Vervaet[6], P. Muñoz[7], G. Garcia[8], J.L. Tarterra[9], B. Fite[9], A. Urtzainki[10], M.C. Vidal[11], E. Sáez Rojo[3], L. Vega Argomaniz[2], A. Civantos[4] and V. Zurdo[3]

*Address all correspondence to: mluengo@icmm.csic.es

1 Polytechnic University of Madrid, Spain

2 Institute of Catalysis and Petroleochemistry, CSIC, Spain

3 Institute of Materials Science of Madrid, CSIC, Spain

4 Institute of Biofunctional Studies, Complutense University of Madrid, Spain

5 INSIGNEO Institute for in silico Medicine, Sheffield, UK

6 Laboratory of Pharmaceutical Technology, University of Ghent, Ghent, Belgium

7 Destilerias Muñoz Galvez S.A., Murcia, Spain

8 Acesur-Tarancon, Carretera Madrid-Valencia, Cuenca, Spain

9 Mahou-San Miguel S.A. Lleida, Spain

10 Createch Medical S. L., Pol. Kurutz Gain, Mendaro (Guipúzcoa), Spain

11 Maicerías Españolas, S.A. (DACSA), Spain

References

[1] 11th International Conference on Renewable Resources & Biorefineries 2015: Conference Proceedings, 3-5 June, 2015, York, United Kingdom.

[2] WasteEng 2014 - 5th International Conference on Engineering for Waste and Biomass Valorisation. August 25-28, 2014. Rio de Janeiro, Brazil.

[3] Braungart M, McDonough W. Editors. Cradle to Cradle: Remaking the Way We Make Things: North Point Press; 2002. ISBN-10: 0865475873

[4] Mirabella N, Castellani V, Serenella S. Current options for the valorization of food manufacturing waste: a review. Journal of Cleaner Production 2014; 65: 28-41.

[5] Ranade VV, Bhandary VM. Editors. Industrial Wastewater Treatment, Recycling and Reuse. Elsevier; 2014. ISBN: 978-0-08-099968-5

[6] Distilleries Muñoz Galvez (DMG). http://dmg.es/ (accessed 24 September 2014).

[7] Project CDTI IDI-20091139 (2010-2011) "Systems for recovery of aquoeous effluents with contaminants valorization and reduction of wastes. Total oxidation of contaminants in effluents through the use of Ecomaterials".

[8] Yates M, Martín-Luengo MA, Vega Argomaniz L, Nogales Velasco S. Design of activated carbon–clay composites for effluent decontamination. Microporous and Mesoporous Materials 2012; 154: 87-92.

[9] Yates M, Blanco J, Martín-Luengo MA, Martin MP. The dynamic adsorption behaviour of volatile organic compounds on activated carbon honeycomb monoliths. Studies in Surface Science and Catalysis 2002; 144: 569-576.

[10] Mahou San Miguel, Spain. http;//mahou-sanmiguel.com. (accessed 24 September 2014).

[11] Yates M, Martín-Luengo MA, Casal Piga B. Patent "Preparation of biocompatible materials from beer production residues and their uses". P200803331.

[12] Maicerías Españolas, S.A. DACSA, Spain. http://dacsa.com. (accessed 24 September 2014).

[13] Martín-Luengo MA, Yates M, Plou F, Lozano R, Saez E, Martínez Serrano AM, Vega L, Medina L, Zurdo V, Ramos M, Valero F, Benaiges MD. Patent "Procedure for production of an immobilized enzyme on a support derived from agroindustrial residues". P201330114.

[14] Martin-Luengo MA, Yates M, Diaz M. Renewable Fine Chemicals from rice and citric subproducts. Ecomaterials. Applied Catalysis B: Environmntal 2011; 106: 488-493.

[15] Visa M, Chelaru AM. Hydrothermally modified fly ash for heavy metals and dyes removal in advanced wastewater treatment. Renewable Energy Systems and Recycling. Applied Surface Science 2014; 303: 14-22.

[16] Yates M, Blanco J, Martin-Luengo MA, Martin MP. Vapour adsorption capacity of controlled porosity honeycomb monoliths. Microporous and Mesoporous Materials 2003; 65(2-3): 219-231.

[17] Martín-Luengo MA, Yates M, Martínez Domingo MJ, Casal B, Iglesias M, Esteban M, Ruiz-Hitzky E. Synthesis of *p*-cymene from limonene, a renewable feedstock. Applied Catalysis B: Environmental 2008; 81: 218-224.

[18] ISO 6060:1989. Water quality Determination of the chemical oxygen demand. http://iso.org/iso/catalogue_detail.htm?csnumber=12260

[19] Ali I, Asim M, Khan TA. Low cost adsorbents for the removal of organic pollutants from wastewater. Review. Journal of Environmental Management 2012; 113: 170-183.

[20] Souza J, Souza PMTG, de Souza PP, Sangiorge DL, Pasa VMD, Oliveira LCA. Production of compounds to be used as fuel additive: Glycerol conversion using Nb-doped MgAl mixed oxide. Catalysis Today 2013; 213: 65-72.

[21] Martin A, Richter M. Oligomerization of glycerol - a critical review. European Journal of Lipid Science and Technology 2011; 113: 100–117.

[22] Project CDTI IDI-20111090 (IDI-20121298) (2012-2014) "Valorization of glycerine subproduct of biodiesel production. Comparison of conventional catalysts and Ecomaterials from own agriresidues".

[23] Behr A, Eilting J, Irawadi K, Leschinski J, Lindner F. Improved utilisation of renewable resources: New important derivatives of glycerol. Green Chemistry 2008; 10: 13-30.

[24] Bagheri S, Julkapli NM, Yehye WA. Catalytic conversion of biodiesel derived raw glycerol to value added products. Review Article. Renewable and Sustainable Energy Reviews 2015; 41: 113-127.

[25] Barrault J, Clacens JM, Pouilloux Y. Selective Oligomerization of Glycerol over Mesoporous Catalysts. Topics in Catalysis 2004; 27: 137–142.

[26] Ruppert AM, Meeldijk JD, Kuipers BWM, Erné BH, Weckhuysen BM. Glycerol Etherification over Highly Active CaO-Based Materials: New Mechanistic Aspects and Related Colloidal Particle Formation. Chemistry- A European Journal 2008; 14(7): 2016-2024.

[27] Barrault J, Jerome F. Design of new solid catalysts for the selective conversion of glicerol. European Journal of Lipid Science and Technology 2008; 110(9): 825-830.

[28] Cassel S, Debaig C, Benvegnu T, Chaimbault P, Lafosse M, Plusquellec D, Rollin P. Original Synthesis of Linear, Branched and Cyclic Oligoglycerol Standards. European Journal of Organic Chemistry 2001; 5: 875–896.

[29] Perozo Rondon E. Basicity in catalysis. A contribution to Green Chemistry. PhD thesis. UNED Madrid; 2008.

[30] Medeiros MA, Araujo MH, Augusti R, Oliveira LCA, LagoRM. Acid-catalyzed oligomerization of glycerol investigated by electrospray ionization mass spectrometry. Journal of the Brazilian Chemical Society 2009; 20: 1667-1673.

[31] Oliveira LCA, Portilho MF, Silva AC, Taroco HA, Souza PP. Modified niobia as a bifunctional catalyst for simultaneous dehydration and oxidation of glycerol. Applied Catalysis B: Environmental 2012; 117–118: 29-35.

[32] Gholami Z, Abdullah AZ, Lee KT. Glycerol etherification to polyglycerols using a Ca_1+xAl_1–xLa_xO_3 composite catalyst in a solventless medium. Journal of the Taiwan Institute of Chemical Engineers 2013; 44: 117-122.

[33] Gil S, Marchena M, Sánchez-Silva L, Romero A, Sánchez P, Valverde JL. Effect of the operation conditions on the selective oxidation of glycerol with catalysts based on

Au supported on carbonaceous materials. Chemical Engineering Journal 2011; 178: 423-435.

[34] Sivaiah MV. Recent developments in acid and base-catalyzed etherification of glycerol to polyglycerols. Review. Catalysis Today 2012; 198: 305-313.

[35] Martín-Luengo MA, Yates M, Ramos M, Martínez AM, Sáez E, Martin Aranda RM, Sanz J, Gil L. Patent "Procedure to obtain multifunctional and renewable materials from sunflower oil production residues" P201130303.

[36] Calero-Rueda O, Barba V, Rodríguez E, Plou FJ, Martínez AT, Martínez MJ. Study of a sterol esterase secreted by Ophiostoma piceae: Sequence, model and biochemical properties. Biochimica et Biophysica Acta (BBA) - Proteins and Proteomics 2009; 194(7): 1099-1106.

[37] Barba V, Plou FJ, Calero-Rueda O, Martínez AT, Martínez MJ. Native and recombinant sterol esterases from Ophiostoma piceae: enzymes with high biotechnological potential. New Biotechnology 2009; 25: S136.

[38] Alcalde M, Ferrer M, Plou FJ, Ballesteros A. Environmental biocatalysis: from remediation with enzymes to novel green processesReview Article. Trends in Biotechnology 2006; 24(6): 281-7.

[39] Hu Q, Tan Z, Liu Y, Tao J, Cai Y, Zhang M, *et al.* Effect of crystallinity of calcium phosphate nanoparticles on adhesion, proliferation, and differentiation of bone marrow mesenchymal stem cells. Journal of Materials Chemistry 2007; 17: 4690-8.

[40] Polo-Corrales L, Latorre-Esteves M, Ramirez-Vick JE. Scaffold design for bone regeneration. Journal of Nanoscience and Nanotechnology 2014; 14: 15-56.

[41] Puppi D, Detta N, Piras AM, Chiellini F, Clarke DA, Reilly GC, *et al.* Development of Electrospun Three-arm Star Poly (ε-caprolactone) Meshes for Tissue Engineering Applications. Macromolecular Bioscience 2010; 10; 887-897.

[42] Hou Q, De Bank PA, Shakesheff KM. Injectable scaffolds for tissue regeneration. Journal of Materials Chemistry 2004; 14: 1915-1923.

[43] Bandyopadhyay A, Bernard S, Xue W, Bose S. Calcium Phosphate-Based Resorbable Ceramics: Influence of MgO, ZnO, and SiO_2 Dopants. Journal of the American Ceramic Society 2006; 89: 2675-2688.

[44] Pietak AM, Reid JW, Stott MJ, Sayer M. Silicon substitution in the calcium phosphate bioceramics.Biomaterials, 2007; 28: 4023-4032.

[45] Kim EJ, Bu SY, Sung MK, Choi MK. Effects of Silicon on Osteoblast Activity and Bone Mineralization of MC3T3-E1 Cells. Biological Trace Elements Research 2013; 152: 105-112.

[46] Zhao SF, Jiang QH, Peel S, Wang XX, He FM. Effects of magnesium-substituted nanohydroxyapatite coating on implant osseointegration. Clinical Oral Implants Research 2011; 24: 34-41.

[47] Wei J, Jia J, Wu F, Wei S, Zhou H, Zhang H, *et al*. Hierarchically microporous/macroporous scaffold of magnesium-calcium phosphate for bone tissue regeneration. Biomaterials 2010; 31: 1260-9.

[48] Martin-Luengo MA, Yates M, Ramos M, Saez Rojo E, Martinez Serrano AM, Gonzalez Gil, L. Biomaterials from beer manufacture waste for bone growth scaffolds. Green Chemistry Letters and Reviews 2011; 4: 229-233.

[49] Martin-Luengo MA, Yates M, Ramos M, Plou F, Salgado JL, Lacomba JL, Reilly G, Vervaet C. Sustainable materials and biorefinery chemicals from agriwastes. In:.V. Abrol V., Sharma P. in Resource management for sustainable agriculture: Intech: 2012, p19-38. Available from http://www.intechopen.com/books/resource-management-for-sustainable-agriculture

[50] Saez Rojo E, Ramos M, Yates M, Martin-Luengo MA, Martínez Serrano MA, Civantos A, *et al.* Preparation, characterization and in vitro osteoblast growth of waste-derived biomaterials. Royal Society Advances 2014; 4: 12630-9. Project INNPACTO IPT-2011-1935-310000 (2011-2014).

[51] Wang D, Christensen K, Chawla K, Xiao GZ, Krebsbach PH, Franceschi RH. Isolation and Characterization of MC3T3-E1 Preosteoblast Subclones with Distinct In Vitro and In Vivo Differentiation/Mineralization Potential. Journal of Bone Mineral Research 1999; 14: 893-899.

[52] Quarles LD, Yohay DA, Lever LW, Caton R, Wenstrup RJ. Distinct proliferative and differentiated stages of murine MC3T3-E1 cells in culture: An *in vitro* model of osteoblast development. Journal of Bone and Mineral Research, 1992; 7: 683-692.

[53] Puleo D, Preston K, Shaffer J, Bizios R. Examination of osteoblastorthopaedic biomaterial interactions using molecular techniques. Biomaterials 1993; 14: 111-4.

[54] Valerio P, Pereira MM, Goes AM, Leite MF. The effect of ionic products from bioactive glass dissolution on osteoblast proliferation and collagen. Biomaterials 2004; 25: 2941-8.

[55] Koh YH, Bae CJ, Sun JJ, Jun IK, Kim HE. Macrochanneled poly (◎-caprolactone)/ hydroxyapatite scaffold by combination of bi-axial machining and lamination. Journal of Materials Science: Materials in Medicine 2006; 17: 773-784.

[56] Budiraharjo R, Neoh KG, Kang ET. Hydroxyapatite-coated carboxymethyl chitosan scaffolds for promoting osteoblast and stem cell differentiation. Journal of Colloid and Interface Science 2012; 366: 224-232.

[57] Kannan S, Vieira SI, Olhero SM, Torres PM, Pina S, Silva OAC et al. Synthesis, mechanical and biological characterization of ionic doped carbonated hydroxyapatite/β-tricalcium phosphate mixtures. Acta Biomaterialia 2011; 7: 1835-1843.

[58] Fratzl-Zelman N, Fratzl P, Hörandner H, Grabner B, Varga F, Ellinger A, *et al.* Matrix mineralization in MC3T3-E1 cell cultures initiated by β-glycerophosphate pulse. Bone 1998; 23; 511-520.

[59] Sudo H, Kodama HA, Amagai Y, Yamamoto S, Kasai S. In vitro differentiation and calcification in a new clonal osteogenic cell line derived from newborn mouse calvaria. Journal of Cell Biology 1983; 96: 191-8.

[60] Longley DB, Harkin DP, Johnston PG. 5-fluorouracil: mechanisms of action and clinical strategies. Nature Reviews 2003; 3(5): 330–8.

[61] Brayfield, A, ed. (13 December 2013). "Fluorouracil". Martindale: The Complete Drug Reference. Pharmaceutical Press. "WHO Model List of EssentialMedicines". World Health Organization. October 2013.

[62] Wang Z, Wang E, Gao L, Xu L. Synthesis and properties of Mg2Al layered double hydroxides containing 5-fluorouracil. Journal of Solid State Chemistry 2005; 178: 736–741.

[63] Diasio RB, Harris BE. Clinical pharmacology of 5-fluorouracil. Clinical Pharmacokinetics 1989; 16: 215-37.

[64] Roik NV, Belyakova LA. Chemical design of pH-sensitive nanovalves on the outer surface of mesoporous silicas for controlled storage and release of aromatic amino acid. Journal of Solid State Chemistry 2014; 215: 284-291

[65] Santos MA Martins RP, Franke MM, Almeida MEV. Calcium phosphate granules for use as a 5-Fluorouracil delivery system Ceramics International 2009; 35:1587-1594.

[66] Datt A, Burns EA, Dhuna NA, Larsen SC. Loading and release of 5-fluorouracil from HY zeolites with varying SiO_2/Al_2O_3 ratios. Microporous and Mesoporous Materials 2013; 167: 182-187.

[67] Cosijns A, Vervaet C, Luyten J, Mullens S, Siepmann F, Van Hoorebeke L, Masschaele B, Cnudde V, Remon JP. Porous hydroxyapatite tablets as carriers for low-dosed drugs. European Journal of Pharmaceutics and Biopharmaceutics 2007; 67: 498-506.